THE M.
ARITHMETIC

By E.A. Sutherland

TABLE OF CONTENTS

	A NOTE FROM THE PUBLISHER	5
	PREFACE	7
Lesson 1	ADDITION & SUBTRACTION	10
Lesson 2	MULTIPLICATION & DIVISION	15
Lesson 3	MEASUREMENTS	19
Lesson 4	PROBLEMS IN NATURE	23
Lesson 5	LIQUID MEASURE	27
Lesson 6	MISCELLANEOUS PROBLEMS	31
Lesson 7	OBSERVATION PROBLEMS	35
	REVIEW 1	38
Lesson 8	TIME & MEASUREMENT PROBLEMS	42
Lesson 9	MEASUREMENTS	46
Lesson 10	PROBLEMS IN PHYSIOLOGY	50
Lesson 11	PROBLEMS IN WEIGHT	54
Lesson 12	DRY MEASURE	59
Lesson 13	PROBLEMS IN PAPERING	64
	REVIEW 2	65
Lesson 14	AREA OF TRIANGLE	70
Lesson 15	PROBLEMS IN BREATHING & TIME	75
Lesson 16	TEMPERATURE PROBLEMS	79
Lesson 17	TIME & WEIGHT PROBLEMS	84
	REVIEW 3	89
Lesson 18	AREAS	95
Lesson 19	TEMPERATURE PROBLEMS	102
	REVIEW 4	106
Lesson 20	PROBLEMS IN CHRONOLOGY	108
Lesson 21	PROBLEMS IN CUBES	114
Lesson 21	LAND PROBLEMS	118
	REVIEW 5	123

Lesson 23 FRUIT-PLANTING PROBLEMS 127

Lesson 24 FRACTIONS & TONS 132

Lesson 25 PROBLEMS IN CHRONOLOGY.................139

REVIEW 6 143

Lesson 26 PROBLEMS IN DECIMALS...................... 149

Lesson 27 PROBLEMS IN BOARD MEASURE..........154

Lesson 28 BOARD MEASURE 160

Lesson 29 THE ERECTION OF A SMALL SCHOOL ......

BUILDING... 165

Lesson 30 PERCENTAGE...173

Lesson 31 PROBLEMS FROM THE FARM 179

REVIEW 7..185

Lesson 32 FRACTIONS 191

Lesson 33 SCIENCE PROBLEMS 196

Lesson 34 INTEREST.. 202

REVIEW 8 207

Lesson 35 NUTRITIVE VALUE OF FOODS 213

Lesson 36 UNITED STATES MONEY 220

Lesson 37 LIQUID MEASURE..................................221

Lesson 38 DRY MEASURE 222

Lesson 39 AVOIRDUPOIS WEIGHT 223

Lesson 40 LINEAR MEASURE 225

Lesson 41 SQUARE MEASURE 226

Lesson 42 CUBIC MEASURE.................................. 227

Lesson 43 TIME 229

MULTIPLICATION TABLE 231

APPENDIX 233

A NOTE FROM
THE PUBLISHER

"But when will I ever use this in real life?" has been the plea of students in math class since math class began. Teachers try their best to make math relevant, but, unfortunately, as traditionally taught, math can seem quite abstract and unrelated-to-real life.

Math is extremely practical; however, educational research tells us that children will learn this subject best when it is taught in a practical way. This classic math book, written by educational pioneer E.A. Sutherland, does just that. The usual abstract and dull math concepts come alive when seen in light of their use in real, everyday life.

This math book is not just another math textbook. Actually, it's not a textbook at all, neither should it be used as one. It is much better than a textbook, and can even replace textbooks when used correctly. As Sutherland explains in his introduction, this book is designed to assist the parent/teacher as they help their child learn math.

This book is for all ages. While young children should not be asked to sit down and do the exercises, parents will find ideas in the book for teaching math through everyday life. Pages ten and eleven, for example, contain simple math learned by counting petals of a flower or toes on a cat. Young children will love this real-life way of learning math. Older youth may appreciate studying through the book themselves and learning math in a conceptual and practical manner. Families, even, can enjoy going through the book together with all ages learning something appropriate to their understanding.

A wide range of subjects are also explored in this book, expanding the scope of learning beyond math alone. Math will come alive through the topics of science, business, agriculture, home construction, Bible chronology, country-living, history, health, and more.

We would encourage parents to use this special volume in a thinking manner. Do not merely hand the book to your child and treat it as another textbook. Use it for your own education in making math a topic useful to real life. Not every problem and exercise must be completed, drills and exercises should be used sparingly, and special emphasis should be given to learning the concepts rather than only the specific examples contained in this book. Used in this manner, your child will enjoy learning math, and this little book can take the place of many volumes of traditional math textbooks, leaving more time to focus on the most important subject of all—developing a character for eternity.

— The Publishers, A Thinking Generation Ministries

PREFACE

It is not because the number of textbooks and arithmetics is limited that the present outline is offered, but because of the recognized lack, in those now in use, of problems which deal with numbers and subject-matter in a practical way, and which are adapted to the needs of children and youth.

The Conference on Mathematics appointed by the "Committee of Ten"* recommended "that the course in arithmetic be at the same time abridged and enriched; abridged by omitting entirely those subjects which *perplex* and *exhaust* the pupil, without affording, any really valuable mental discipline; and enriched by *a greater number of exercises* in simple calculations, and in the solution of *concrete problems.*

This shows a demand for a reformation. Why is a change necessary? "In the arithmetics of the past are mirrored the methods of instruction that prevailed in that time," and true it is that in the history of arithmetics may be traced the story of national progress. This close relationship between the arithmetic and the people is especially noticeable in Roman history, and it is because of the mold which was given to the subject by that nation, and which has been too closely adhered to since, that makes the change now necessary. What was the Roman mold? — In her greed for worldly gain, Rome stamped everything with the impress of commercialism. Her arithmetic, with all things else, received this seal of the state. Most of the textbooks of today bear heavily along this same line, to the exclusion of other and equally practical material.

Horace Mann, with the aid of Pliny E. Chase, about 1840, attempted a reform in the arithmetical work of the

* The National Educational Association that met at Saratoga in July, 1892, appointed the "Committee of Ten."

schools. The author thus expressed the plan of the work which he issued: "It derives its examples from biography, geography, phrenology, and history; from education, finance, commercial, and civil statistics; from the laws of light and electricity, of sound and motion, of chemistry and astronomy, and others of the exact sciences. Trades, handicrafts, and whatever pertain to the useful arts, are laid under contribution, and are made to supply appropriate elements in the questions on which the youthful learner may exercise his arithmetical faculties."

It is refreshing to find that at the most critical period in the educational history of the United States, an arithmetic such as that of Horace Mann's should have been published. Today the work begun by Horace Mann must be carried to completion, and no apology is necessary for the present volume, which, though elementary and mental, deals with practical subjects and matters with which every child is, or ought to be, familiar. That there is need of a reform is shown by the character of the subject-matter in a large number of the arithmetics now in use. A casual glance is all that is necessary to show that the pupil is required to spend no little time in the solution of problems dealing with wines, tobacco, false measures, etc., which keep before the mind a side of life which is not conducive to honesty and uprightness of character. The spirit of commercialism adheres firmly to many so-called reforms in arithmetic, and in an attempt to improve upon the subject-matter of the problems, wines and tobacco are exchanged for pies and candies, etc.

Science offers a broad field for wholesome practical problems, but the work in this sphere is usually one sided, avoiding altogether the commercial problems. It has been the aim in this arithmetic to avoid extremes: the value of science study is recognized, likewise the need of accuracy in business forms.

The recent suggestions of some prominent educators will explain why some subjects are omitted which appear in other textbooks. They recommend that below the fifth grade —

"1. There should be no long division with divisors of more than two figures.

"2. Work in fractions should be made oral.

"3. The greatest common divisor should be omitted as a separate topic.

"4. Longitude and time should be omitted.

"5. Little attention should be given to problems in interest."

Mothers have long looked for a book in arithmetic which would guide them in giving the child its first lessons. The present volume is for them. They need not confine themselves to the problems given, but these are suggestive of arithmetical work to be done in the home as well as in the schoolroom.

The value of arithmetic depends upon the accuracy and speed which are developed. These qualities come as the result of thorough drills. The fundamental principles of arithmetic must become tools in the hands of the pupil, which can be used without blundering.

Attention is called to such features as —

1. The subject-matter of the problems.
2. The frequent drills.
3. The schemes for insuring accuracy and speed.
4. The emphasis laid on mental work for beginners.
5. The amount of actual work required, such as drawing, paper cutting, measuring, etc.
6. The summary of tables, and valuable information in the last pages.

The science of numbers is treated in a rigorous method, and at the same time the learner is acquiring useful and valuable information.

NOTE
See "Suggestions to Teachers" in Appendix.

ADDITION & SUBTRACTION

LESSON 1

1. How many stars in the handle of the Big Dipper? How many in the bowl? Which are *pointers*?

"The heavens declare the glory of God."

2. The stars in the handle added to the 2 pointers make how many stars?

3. The stars in the handle added to the 4 stars in the bowl make _____ stars.

4. The *pointers* point to the North star. Add the North star to the group in the Dipper; how many stars in all?

5. Think of the Dipper when you read this:

    $$3 + 4 + 1 = \underline{\qquad}$$

    What does each figure mean?

> **NOTE**
> The teacher should direct the attention of the pupils to the northern heavens, familiarizing them with the Dipper and other constellations.

"Consider the lilies."

6. Count the flower leaves of the lily. How many pollen boxes has the lily? How many stems to support the pollen boxes are there in each lily?

7. The lily faded and lost 2 flower leaves; how many were left?

   Six less two equals _____.

   6 – 2 = _____

8. Two blossomed on the same plant. Counting the flower leaves on both, how many were there?

   Six plus six equals _____.

   6 + 6 = _____

9. A wild rose grew beside the lily. Count its flower leaves. How many more flower leaves has the lily than has the rose?

   The difference between 6 and 5 is _____.

   6 – 5 = _____

10. Count the flower leaves on the 2 lilies and the 1 rose.

11. How many more flower leaves have 2 wild roses than has 1 lily?

    5 plus 5 less 6 = _____

12. How many more flower leaves have 2 lilies than 2 wild roses ?

    6 plus 6 less 10 = _____

13. Count the bones in your index finger. Count the bones in your thumb. How many bones in 1 finger and 1 thumb?

    3 plus 2 = _____

14. How many bones in 2 thumbs? How many more bones in 1 finger and 1 thumb than in 2 thumbs?

    2 plus 2 = _____

    3 plus 2 less 4 = _____

15. How many more bones in 2 fingers than in 1 thumb? How many more in 2 fingers than in 2 thumbs?

    3 plus 3 less 2 = _____

    3 plus 3 less 4 = _____

16. Count the toes on the feet of a cat. How many are there on the 2 fore feet? How many on both rear feet?

17. How many more toes on 2 fore feet than on 1 rear foot? How many more on the 2 fore feet than on both rear feet?

18. How many toes on 1 fore foot and 1 rear foot?

19. How many hoofs has the cow on all 4 feet?

20. How many more toes on the 2 rear feet of a cat than on the 2 rear feet of a cow?

21. Make two problems using the numbers 4 and 2 in each.

 DRILL

1. Three stars plus 2 stars equals _____ stars.
2. Three stars plus 4 stars equals _____ stars.
3. Two thumb bones plus 2 bones equals _____ bones.
4. Three finger bones plus 3 finger bones equals _____ fingerbones.
5. Three stars plus 5 stars equals _____ stars.
6. Six flower leaves minus 3 flower leaves equals _____ flower leaves.
7. Five flower leaves plus 5 flower leaves equals _____ flower leaves.
8. Three stars plus 1 star plus 4 stars equals _____ stars.
9. Two hoofs plus 2 hoofs plus 2 hoofs plus 2 hoofs equals _____ hoofs. Four 2's = _____.
10. There are _____ stars in the handle of the dipper.

SIGHT EXERCISES

✓ 3 + 2 = _____   3 + 4 = _____   3 + 3 = _____

6 – 3 = _____   6 – 4 = _____   6 – 2 = _____

4 + 2 = _____   6 + 5 = _____   6 – 5 = _____

✓ Add:

2	3	4	6	3	4	5
+2	+3	+2	+5	+2	+4	+4

✓ Subtract:

6	6	10	6	6	3	4	5
– 4	– 2	– 6	– 3	– 5	– 2	– 2	– 4

✓ Two 2's = _____ Two 6's = _____ Two 3's = _____

Two 5's = _____ Two 4's = _____ Two 1's = _____

MULTIPLICATION & DIVISION

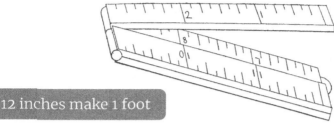

12 inches make 1 foot

1. How many inches in a foot rule? How many inches in one half a foot rule?

   ½ of 12 = _____   12 divided by 2 = _____   12:2 = ___

2. I have a foot rule which folds in 4 sections. Count the inches in each section. How many inches in two sections?

   3 + 3 = _____   Two times 3 = _____   2 x 3 = ___

3. How long is the book which requires 3 sections of the foot rule to measure it?

   3 + 3 + 3 = _____   Three times 3 = _____   3 x 3 = ___

4. Measure your reader. How many more inches is it in length than in width?

5. How many more inches in the length than in the width of your hand?

6. Measure your arm. Give the result in feet and inches.

7. If a child's arm is once the length of the foot rule plus the length of 1 section of the rule, how many inches long is it?

8. With a string measure the distance from the tip of your nose to the end of your right hand with the arm

extended. How many times the length of the foot rule is the string?

9. How many inches is it around your wrist? What is the distance around 2 wrists?

10. How much longer is it from your wrist to your elbow than it is around your wrist ?

11. Measure your arm at the largest part. How much larger is it when the elbow is bent than when the arm is straight? Find the difference between this arm measure and the wrist measure.

12. How many inches wide must your coat sleeve be in order to slip over the closed fist?

13. A glove fitter measures the hand of his customer across the knuckles. How many inches around your hand?

14. In the average person, twice the length around the wrist equals the distance around the neck. Is this true in your case? What is the size of the collar which fits you?

15. Mark your height on the wall. How many times the length of the foot rule are you in height?

16. When a well-formed man extends both arms, the distance from tip to tip of his fingers is equal to his height. Are these two measurements the same in your case?

17. Measure your body over the floating ribs when the lungs are not filled. Take a deep breath. What is the difference in inches in the two measurements?

18. How many times the length of the foot rule is the distance from your knee to the sole of your foot?

19. Take other measurements and make 2 original problems.

1. Two and 2 are _____. $2 + 2 =$ ___
2. Six less 2 are _____. $6 - 2 =$ ___
3. Two times 4 are _____. $2 \times 4 =$ ___
4. How many 2's are there in 8? _____. $8 : 2 =$ ___
5. One half of 6 is _____.   ½ of $6 =$ ___
6. Three and 2 are _____. $3 + 2 =$ ___
7. Eight less 6 are _____. $8 - 6 =$ ___
8. Two times 3 are _____. $2 \times 3 =$ ___
9. How many 2's are there in 6? _____. $6 : 2 =$ ___
10. One half of 8 is _____.   ½ of $8 =$ ___
11. Three inches and 2 inches are _____.
12. Five inches less 3 inches are _____.
13. How many inches are there in ½ of a foot? _____.
14. One half of a foot and two inches are _____ inches.
15. If my book is ½ of a foot long, and yours is 2 inches shorter; how long is yours? _____.
16. How many inches are there in ¼ of a foot? _____.
17. One fourth of a foot and 4 inches are _____ inches.
18. One half of a foot and ¼ of a foot equal ___ inches.

✓ Add:

2	3	2	3	4	3	2	2
+ 2	+ 3	+ 3	+ 4	+ 4	+ 5	+ 4	+ 5

2	4	3	5	2	2	2	3
+ 6	+ 5	+ 6	+ 5	+ 7	+ 8	+ 9	+ 8

4	2	3	5	5	4	3	4
+ 6	+ 10	+ 9	+ 6	+ 7	+ 8	+ 7	+ 7

✓ 2 x 2 = ___ 　3 x 2 = ___ 　3 x 3 = ___ 　3 x 4 = ___

　2 x 3 = ___ 　2 x 4 = ___ 　2 x 5 = ___ 　2 x 6 = ___

✓ Subtract:

7	8	6	7	9	8	9	8
– 5	– 4	– 5	– 3	– 7	– 5	– 5	– 3

10	9	12	9	10	11	12	10
– 5	– 3	– 6	– 3	– 6	– 7	– 7	– 2

✓ 4 : 2 = ___ 　6 : 2 = ___ 　10 : 2 = ___ 　12 : 2 = ___

　9 : 3 = ___ 　8 : 2 = ___ 　12 : 4 = ___ 　12 : 3 = ___

MEASUREMENTS

1. Cut out a piece of paper 2 inches wide and 4 inches long. Fold the paper so that the short edges will meet; crease and open. Fold the paper so that the short edges will meet the crease in the center; crease and open. Fold the paper so that the long edges will meet; crease and open.

2. How many square inches are there on your paper? Or, what is the area of the paper in square inches?

3. Four square inches are what part of the area?

4. Two square inches are what part of the area?

5. I have a square containing 9 square inches. How long and how wide is it ?

6. Cut out a piece of paper containing just 9 square inches, but do not have it square. How long and how wide is it?

7. Cut a piece of paper just large enough to contain 12 square inches, having one side 3 inches long. How long must the other side be?

8. How many rows of 3 square inches are there in the square containing 9 square inches?

9. How many rows of 3 square inches are there in the rectangle containing 12 square inches?

10. Cut out another piece of paper having an area of 12 square inches. Let one side be 2 inches long; how long must the other side be? How many rows of two square inches are there in this paper?

11. A table top is 3 feet wide and 4 feet long. How many square feet in its area?

12. Prove the last example with a piece of paper, letting 1 inch represent a foot.

13. Mabel is 10 years old and has been in school 2 years. How old was she when she started to school?

14. If her 4-year-old brother begins school at the age she did, how long will it be before he can start?

15. John and Peter together wrote 7 books in the New Testament. Peter wrote 2. How many did John write?

16. How many books of the Bible did Moses write? Counting all together, how many books did Moses, John, and Peter write?

17. Paul wrote two more books than Moses, John, and Peter wrote; how many books did he write?

18. How many more books did Paul write than did John?

 **DRILL**

1. Two 3's are _____.
2. Three 3's are _____.
3. Nine less 4 are _____.
4. Three times 3 are _____.
5. One third of 9 is _____.
6. Eight and 2 are _____.
7. Eight less 5 are _____.
8. Five is what part of 10? _____.
9. Two times 5 are _____.
10. Fourteen less 4 are _____.
11. Seven less 2 are _____.
12. Five and 5 and 2 are _____.
13. Twelve and 2 are _____.
14. Fourteen less 5 are _____.
15. Two times 4 are _____.
16. Four is what part of 8? _____.
17. Two is what part of 8? _____.
18. Six and 4 are _____.
19. Two 3's = _____.   Two 4's = _____.
    Two 5's = _____.   Two 6's = _____.
    Two 7's = _____.
20. Which is more:
    2 times 6   or   3 times 4?
21. John wrote _____ books.
    Peter wrote _____ books.
22. Moses wrote _____ books.
    _____ books were written by Paul.

 **SIGHT EXERCISES**

✓ Add:

3	4	4	5	3	3	4	5
+ 5	+ 6	+ 5	+ 6	+ 8	+ 6	+ 8	+ 7

2	1	2	1	2	3	3	2
3	2	4	5	5	3	4	5
+ 5	+ 6	+ 6	+ 5	+ 6	+ 6	+ 4	+ 7

✓ 4 x 2 = ___   5 x 2 = ___   6 x 2 = ___   2 x 6 = ___

  3 x 3 = ___   4 x 3 = ___   2 x 7 = ___   7 x 2 = ___

✓ 4 and what number = 6? ___   ✓ 6 – 3 = ___

  5 and what number = 10? ___   7 – 2 = ___

  4 and what number = 8? ___   11 – 5 = ___

  7 and what number = 10? ___   12 – 7 = ___

  6 and what number = 12? ___   13 – 3 = ___

✓ 4 + 2 – 3 = ___   6 + 4 – 3 = ___   9 + 3 – 4 = ___

  5 + 3 – 2 = ___   7 + 3 – 5 = ___   8 + 5 – 5 = ___

  6 + 2 – 3 = ___   7 + 5 – 2 = ___   10 + 4 – 3 = ___

  7 + 3 – 4 = ___   8 + 3 – 4 = ___   12 – 4 + 2 = ___

✓ One half of 6   = ___   One half of 10   = ___

  One third of 9 = ___   One third of 12 = ___

  One fourth of 8 = ___   One fifth of 10 = ___

  One half of 14 = ___   One sixth of 12 = ___

PROBLEMS IN NATURE

1. The pear has 5 seed-cells. If there are two seeds in each cell, how many seeds has the pear?

2. The apple also has 5 seed-cells. If, in some particular apple, there are 3 seeds in each of 3 cells, and 2 seeds in each of the other cells, how many seeds does the apple have?

3. How many more seeds does the apple have than the pear?

Section of an apple

4. How many seeds has an apple that has 3 seeds in each seed-cell?

5. If an apple has 4 seeds in each of 2 seed cells, and 2 seeds in each of the other cells, how many seeds does it have?

6. How many seeds has an apple that has 4 seeds in each of two cells, 3 seeds in each of two other cells, and 1 seed in the remaining cell?

7. Each flower of the apple and the pear has the same number of petals as there are seed-cells in the fruit. How many petals in one pear blossom and two apple blossoms?

8. Three apple blossoms grew in a cluster. After the wind had blown two petals from each blossom, how many remained?

9. Sometimes instead of saying 12 eggs, we say a _____ eggs.

10. A dozen eggs and 3 eggs are _____ eggs.

11. When eggs are selling for 12 cents a dozen, how much must I pay for $\frac{1}{3}$ of a dozen?

12. If 3 bananas cost 5 cents, how much does $\frac{1}{2}$ of a dozen cost?

13. At 2 for 5 cents, how much does $\frac{1}{2}$ of a dozen bananas cost?

14. What is another name for 12 months?

15. How many seasons are there in the year? Name the seasons. How many months are there in each season? Two seasons are what part of the year? Three seasons are what part of the year?

16. How many disciples did Jesus have? How many were with him on the Mount of Transfiguration? What part of the whole number were with him?

17. How many were absent at that time? What fraction of the whole number were absent?

18. Twelve inches make 1 foot.

    Twelve things make 1 dozen.

    Twelve months make 1 year.

    Jesus had 12 disciples.

    Remember these 4 twelves.

19. Make a problem using the numbers 10 and 5.

    Make problems using *dozen* and *cents*.

 DRILL

1. One and ½ dozen eggs at 10 cents a dozen = _____
2. One half dozen eggs at 16 cents a dozen = _____
3. One dozen + 3 = _____
4. One third of one dozen = _____
5. One fourth of a dozen eggs at 16 cents per dozen = _____
6. If ¼ of a dozen bananas costs 5 cents, one dozen costs _____ cents.
7. How many months in 1 and ⅓ years? _____
8. How many months in 1 and ¼ years? _____
9. How many seasons in 5 years? _____
10. How many months in 6 winters? _____
11. How many inches in 2 feet? _____
12. How many inches in 3 feet? _____
13. There are _____ inches in 1 and ½ feet.
14. Sixteen inches = how many feet? _____
15. Two feet minus 18 inches = _____ inches
16. _____ inches make 1 foot.
17. 12 things make 1 _____.
18. _____ months make 1 year.
19. 10 eggs = 1 dozen minus _____.
20. 15 eggs = 1 dozen plus _____.
21. 18 months = 1 year plus _____ months.

✓ Add:

5	4	10	9	8	8	7
+ 6	+ 5	+ 3	+ 5	+ 9	+ 8	+ 8

6	9	8	5	4	6	7
4	3	7	5	5	6	8
+ 2	+ 2	+ 3	+ 5	+ 1	+ 6	+ 3

✓ 5 x 2 + 6 = ____   ✓ 16 : 2 + 2 = ____   ✓ ½ of 6   = ____

3 x 3 + 2 = ____   15 : 3 + 2 = ____   ⅓ of 6   = ____

3 x 3 + 3 = ____   18 : 2 – 6 = ____   ¼ of 8   = ____

5 x 3 + 1 = ____   12 : 2 + 4 = ____   ½ of 12 = ____

5 x 2 + 7 = ____   12 : 4 + 3 = ____   ⅓ of 12 = ____

3 x 4 + 6 = ____   15 : 5 + 8 = ____   ¼ of 12 = ____

5 x 2 + 10 = ____   9 : 3 + 2 = ____   ⅓ of 9   = ____

6 x 2 + 6 = ____   8 : 4 + 2 = ____   ½ of 14 = ____

6 x 3 + 1 = ____   8 : 4 + 10 = ____

7 x 1 + 7 = ____   9 : 3 + 6 = ____

5 x 3 + 3 = ____   15 : 5 – 2 = ____

4 x 3 – 2 = ____   14 : 2 + 8 = ____

✓ 4 is what part of 8? ____

4 is what part of 12? ____

3 is what part of 9? ____

3 is what part of 12? ____

2 is what part of 6? ____

3 is what part of 6? ____

5 LIQUID MEASURE

For the following problems, use pint, quart, and gallon measures.

1. How many pints does the quart measure hold?
2. How many quarts does the gallon measure hold?
3. How many times can the pint measure be filled from a gallon of water?

4. How many quarts does your water-pail hold?
5. If a 2-gallon water-pail were half full, and twelve of you should each drink one half pint, how many pints would be left in the pail?
6. During the day, but not at meals, every person ought to drink 3 pints of water. At this rate, how many quarts do 4 persons need in 1 day? How many gallons does this make?
7. How many pints of water does it take to fill the normal stomach of an adult?

8. If a cow gives 6 quarts of milk in the morning and 2 gallons at night, how many quarts of milk does she give in one day?

9. If the owner of the cow uses 2 quarts of the milk, and sells the balance each day, how many gallons does he sell?

10. How much does he get for 3 quarts at 5 cents a quart?

11. How much does he receive for a gallon of milk at 5 cents a quart?

12. What would the tithe on the price of 1 gallon of milk be?

> **NOTE**
> Read about tithe in Malachi 3:10 and Leviticus 27:30.

13. What tithe would he pay on 3 gallons?

14. What would his tithe on the milk sold amount to in 1 week?

15. What would be the tithe on the amount sold for 12 days?

16. What would be the tithe on the whole amount of milk for 12 days?

17. What is the difference between the tithe on the whole amount of milk and the tithe on the amount sold for 12 days?

18. When kerosene sells for 12 cents a gallon, how much does a quart cost?

19. At 4 cents a quart, how much does 1 gallon of gasoline cost?

20. _____ pints make 1 quart.

_____ quarts make 1 gallon.

BIBLE MEASURES

1 hin	=	1 gallon
1 firkin	=	9 gallons

1. Read Exodus 29:40, and find the number of quarts of oil and wine used in the sacrifice.
2. Read John 2:6. Provided the water jars used at the marriage feast of Cana held 2 firkins apiece, how many gallons were there in each firkin? How many gallons in a firkin ½ full?

 DRILL

1. Four pints equal _____ quarts.
2. One quart and 3 pints equal _____ pints.
3. Two pints are what part of a gallon? _____
4. From a gallon of milk, 5 pints were sold. How many pints remained? _____
5. One quart is what part of a gallon? _____
6. Six quarts and 2 gallons equal _____ quarts.
7. Fourteen quarts less 2 quarts equal _____ gallons.
8. Two 5's are _____.
9. Three 3's and two 2's are _____.
10. Thirteen less 10 are _____.
11. Five 3's are _____.
12. Five is what part of 15? _____
13. Two 4's and three 2's are _____.
14. Two 4's, two 3's, and 1 are _____.
15. One dozen and 3 are _____.
16. One half of a dozen and 4 are _____.
17. One third of a dozen is _____.
18. One fourth of a year equals _____.
19. Give the group of 12.

✓ Add:

4	3	5	3	4	5	6	5
+ 6	+ 8	+ 6	+ 7	+ 8	+ 9	+ 8	+10

2	3	3	3	2	3	2	2
4	3	2	5	4	3	3	5
+ 6	+ 4	+ 6	+ 7	+ 8	+ 9	+10	+ 8

✓ 3 x 2 + 4 = ____   ✓ 12 : 2 + 3 = ____

3 x 3 + 2 = ____   10 : 5 + 8 = ____

4 x 2 - 3 = ____   12 : 4 + 6 = ____

2 x 5 + 4 = ____   14 : 2 - 3 = ____

2 x 3 + 6 = ____   8 : 2 + 6 = ____

3 x 3 + 5 = ____   9 : 3 + 8 = ____

3 x 4 - 4 = ____   10 : 2 + 9 = ____

4 x 3 + 3 = ____   15 : 5 + 7 = ____

2 x 7 - 3 = ____   12 : 6 + 12 = ____

5 x 2 + 5 = ____   8 : 4 + 7 = ____

3 x 5 - 4 = ____   15 : 3 + 8 = ____

2 x 6 + 3 = ____   15 : 3 - 5 = ____

✓ 4 is what part of 8? ____   ✓ 5 is what part of 10? ____

3 is what part of 9? ____   2 is what part of 12? ____

2 is what part of 10? ____   3 is what part of 15? ____

3 is what part of 12? ____   2 is what part of 14? ____

LESSON 6

MISCELLANEOUS PROBLEMS

1. Mary's mama gave her a dime with which to buy bread. Bread is worth 5 cents a loaf. How many loaves could she buy?

2. Sometimes we can buy bread for 4 cents a loaf. How many cents would the baker have returned to Mary if she had bought two loaves at 4 cents a loaf?

3. How many cents would three loaves have cost at 5 cents a loaf? How much would 4 loaves have cost at 4 cents a loaf?

4. In 3 pounds of ordinary wheat bread, there is 1 pound of water, and the balance is solid food. How many pounds of solid food are there in 6 pounds of wheat bread?

5. One half of ordinary wheat bread is starch. How many pounds of starch may be obtained from 6 pounds of bread? from 8 pounds? from 14 pounds? from 16 pounds?

6. When bread costs 5 cents a pound, how much do you pay for the starch in it?

7. In 14 pounds of ordinary bread, there is 1 pound of gluten. How many more pounds of starch are there in the same quantity?

8. John picked 9 quarts of strawberries, and Harry picked 7 quarts. How many quarts did both together pick?

9. If they were paid 4 cents a gallon for picking berries, how much money did they make?

10. If the canned berries filled 5 2-quart jars, how many quarts were lost in waste and shrinkage?

11. Helen picked 3 quarts of strawberries and sold them for 6 cents a quart, receiving in payment 2 dimes. How many cents did she return to the purchaser?

12. She bought 2 pounds of prunes at 8 cents a pound. How many cents had she left?

> **NOTE**
> The liquid quart is used in the berry problems of this lesson.

13. Complete the following multiplication table, then repeat from memory.

2 x 1 = 2	2 x 5 = ___	2 x 9 = ___
2 x 2 = 4	2 x 6 = ___	2 x 10 = ___
2 x 3 = ___	2 x 7 = ___	2 x 11 = ___
2 x 4 = ___	2 x 8 = ___	2 x 12 = ___

 DRILL

1. How many nickels, or 5-cent pieces, in a dime? ___
2. Ten cents less 8 cents are _____ cents.
3. Three nickels equal _____ cents.
4. One third of 3 is _____.
5. Two thirds of 6 is _____.
6. One half of 14 is _____.
7. Seven less 1 are _____.
8. Nine and 7 are _____.
9. Four 4's are _____.
10. How many 2's are there in 16? _____
11. Sixteen less five 2's are _____.
12. Seven and 6 are _____.

13. Thirteen less 3 = _____.

14. Thirteen less 8 = _____.

15. Eleven less 4 = _____

16. Eleven less 6 = _____

17. How many 3's are there in 15? _____.

18. One third of 15 is _____.

19. How many 4's are there in 16? _____

👓 SIGHT EXERCISES

✓ Add:

2	3	2	3	1	4	3	5
3	3	4	4	4	6	6	5
+ 7	+ 6	+ 6	+ 8	+ 9	+ 3	+ 7	+ 5

1	2	3	2	2	3	3	2
3	3	3	4	5	3	4	5
4	3	3	4	1	3	4	2
+ 5	+ 4	+ 4	+ 4	+ 7	+ 3	+ 4	+ 7

✓ 2 x ___ = 12   ✓ ½ of 14 = ___

3 x ___ = 15   ⅓ of 12 = ___

3 x ___ = 6   ¼ of 12 = ___

4 x ___ = 12   ⅓ of 9 = ___

5 x ___ = 15   ⅕ of 10 = ___

4 x ___ = 8   ⅙ of 12 = ___

3 x ___ = 9   ½ of 16 = ___

2 x ___ = 6   ⅕ of 15 = ___

4 x ___ = 16   ⅓ of 18 = ___

✓ 3 x 3 + 2 - 1   = ___   ✓ 2 x ___ = 14

2 x 6 + 4 - 2   = ___   2 x ___ = 2

3 x 5 - 5 + 2   = ___   2 x ___ = 8

3 x 4 + 3 - 5   = ___   2 x ___ = 16

2 x 7 - 4 + 3   = ___   2 x ___ = 12

2 x 5 + 5 - 7   = ___   2 x ___ = 4

2 x 8 - 6 + 4   = ___   2 x ___ = 10

3 x 6 - 7 + 2   = ___   2 x ___ = 24

3 x 5 + 2 - 8   = ___   2 x ___ = 6

3 x 4 + 6 - 9   = ___   2 x ___ = 18

3 x 6 - 9 + 2   = ___   2 x ___ = 20

4 x 4 + 2 -10   = ___   2 x ___ = 22

✓ Count by 2's from 2 to 18.
Count by 2's from 1 to 17.
Count by 3's from 3 to 18.
Count by 3's from 1 to 16.
Count by 3's from 2 to 17.

> **NOTE**
> Drill thoroughly on the multiplication table.

OBSERVATION PROBLEMS

1. There is the same number of bones in the ear as in the finger. How many bones are there in both ears?

   Two times 3 are _____.

   > These little ear bones are called the hammer, anvil, and stirrup. Why are they given these names?

2. The finger bones are called phalanges. How many phalanges in 3 fingers?

   Three times 3 are _____.

   > The word *phalanges* is the plural of the Greek *phalanx—a line, or a rank, of soldiers*; hence in anatomy the row, or series, of bones in the fingers and toes.

3. How many phalanges are there in the hand?

   Four times 3 are _____.

   Are there the same number of bones in the toes as in the fingers?

4. There are 7 tarsal bones in the ankle. How many are there in both ankles?

   Two times 7 are _____.

   > Tarsal bones—the collection of bones that forms the ankle-joint and instep.

5. There are 4 bones in the leg above the ankle. How many bones in the ankle and leg together?

   Two times 7 plus 4 are _____.

   > What is the longest bone in the body? What is its name?

6. The wrist has 1 more bone than has the ankle. How many carpal bones in both wrists?

> The word *carpal* is from the Greek *karpos*, the wrist.

7. How many bones are there in the hand between the phalanges and the carpal bones? They are called metacarpal bones.

> The word *metacarpal* is from the Greek *meta*—beyond, and *karpos*—the wrist: beyond the wrist.

8. How many bones are there in the arm above the carpal bones, counting the collar bone and the shoulder blade?

9. There are as many bones in the face as there are tarsal bones in both ankles. How many bones are there in the face?

10. The African elephant has 4 nails on each forefoot, but only 3 nails on each rear foot. How many more nails has it on both fore feet than on both rear feet?

    Two times 4 minus 2 times 3 are _____.

11. The Indian elephant has one more hoof on each foot than has the African elephant. How many hoofs has it on both fore feet? How many more on both fore feet than on both rear feet?

    Two times 5 minus 2 times 4 are _____.

12. How many more hoofs has the Indian elephant than has the African elephant?

13. The rhinoceros has 3 toes, encased in hooves, on each foot. How many toes has it on all its feet?

14. The hippopotamus has 4 hoofed toes on each foot. How many has it on all its feet?

15. Make three original problems about animals that you have seen, using the numbers 3, 4, and 5.

16. Complete the following multiplication table and repeat from memory.

3 x 1 = 3   3 x 5 = ___   3 x 9 = ___
3 x 2 = 6   3 x 6 = ___   3 x 10 = ___
3 x 3 = ___   3 x 7 = ___   3 x 11 = ___
3 x 4 = ___   3 x 8 = ___   3 x 12 = ___

REVIEW 1

1. How many days are there in the week? How many of these are working days?

2. How many school days are there in 3 weeks?

3. How many more working days are there than school days in three weeks?

4. Into how many colors may a ray of sunlight be divided?

5. How many steps are there in the musical scale?

6. How many groups of the number 7 can you find in the book of Revelation?

7. Nine months are what part of the year? Name the months of the year.

8. When milk is worth 20 cents a gallon, what is the price of a quart?

9. When bananas are worth 15 cents a dozen, how many may be bought for 5 cents?

10. When they are worth 20 cents a dozen, how many may be bought for 5 cents?

11. There are 6 New England states and 7 Middle Atlantic states. How many states are there in these two groups?

> **NOTE**
> Use a map with all geographical problems.

12. Three of these states have no sea coast. How many of them border on the sea?

13. Eight of them are mountainous. How many have no mountains?

14. Of the 11 Southern states, 4 are west of the Mississippi River. How many are east of it?

15. How many of these states have no sea coast? How many have a sea coast?

16. Six of them are mountainous. How many are not?

17. Find from the map how many of these states border on the Mississippi River; and then without counting, tell how many do not.

18. How many border on the Gulf of Mexico, and how many do not?

19. How many border on the Atlantic Ocean, and how many do not?

20. How many loaves of bread may be bought for 20 cents at 4 cents a loaf?

21. If each loaf weighs a pound, how many pounds of starch do you get?

22. How much gluten do you get in 7 pounds of ordinary bread?

23. About how many pounds of water are there in 9 pounds of bread?

24. How many inches are there in a foot and a half?

25. The lion belongs to the cat family. How many toes has it on all four feet?

26. How many toes has the dog on all four feet?

27. What is the difference in the number of legs of the spider and of the fly? How many toes has a fowl?

28. Cloven-footed animals that chew the cud have a double stomach which has one half as many compartments as the animal has hooves. How many rooms are there in the stomach of a cow?

29. A man worked 12 days at a dollar and a half a day. How much money did he earn?

30. How much was the tithe of it?

31. William bought 3 pounds of 8-penny nails, and in payment gave the storekeeper two dimes. He received in change a nickel and 3 cents; how much did the nails cost? What was the price per pound?

✏ DRILL

1. Fourteen less 8 are _____.
2. Fourteen and 5 are _____.
3. One half of 14 is _____.
4. How many 2's are there in 14? _____
5. Name 3 groups of 14 that we have had in the bones of the human body. _____
6. Seven, 5, and 3 are _____.
7. Four 3's are _____.
8. Five is what part of 20? _____
9. How many nickels are there in 20 cents? _____
10. Ten, 4, and 3 are _____.
11. Twenty cents less 8 cents are _____ cents.
12. A foot and a half of a foot equals _____ inches.
13. Two dimes less 3 cents equals _____ cents.
14. A dime, a nickel, and 3 cents equal _____ cents.
15. Six is what part of 18? _____
16. How many 3's are there in 18? _____
17. Which is the more—four 5's or five 4's? _____
18. Two 5's and three 3's equal _____.
19. Six 3's less four 2's equal _____.

👓 SIGHT EXERCISES

✓ Add:

3	2	3	4	5	3	4	5
4	6	5	2	6	7	6	6
+ 7	+ 7	+ 8	+ 8	+ 7	+ 8	+ 9	+ 9

2	3	2	3	4	3	4	3
3	3	4	3	4	5	5	4
4	6	4	5	4	2	2	3
+ 8	+ 6	+ 8	+ 5	+ 4	+ 6	+ 7	+ 7

✓ DRILL IN MULTIPLYING & ADDING

								3
1	2	3	4	5	6	7	8	9
								x 2

Instructions:

1. Write the numbers from 1 to 9 on the blackboard. Underneath, place a multiplier. Above, place a number to be added to the product.
2. Placing the pointer at 8, the pupil says 16, 19. At 6, he says 12, 15.
3. Change the multiplier to 3.
4. When the teacher points to 6, the pupil says 18, 21, etc.
5. Change the adding figure to 4. Pointing to 5, the reply is 15, 19.
6. Do not in this lesson make combinations above 20.

✓ ¼ of 16 = ___   ½ of 18 = ___

⅓ of 15 = ___   ¼ of 20 = ___

⅙ of 18 = ___   ¹⁄₁₀ of 20 = ___

½ of 20 = ___   ⅕ of 20 = ___

⅓ of 18 = ___

TIME & MEASUREMENT PROBLEMS

> 60 seconds make 1 minute.
> 60 minutes make 1 hour.
> 24 hours make 1 day.

1. How many hours are there from one sunset until the next?

2. How many times does the hour hand of the clock go around from one sunset until the next?

3. How many hours are there from the time you get up in the morning until noon?

4. Show the position of the hands of the clock at half past 5 o'clock in the morning. What are their positions at half past 5 o'clock in the afternoon? At half past 8 o'clock?

5. If you finish your breakfast at 7 o'clock, how many times does the minute hand go around before school time, when school begins at 9 o'clock?

6. How long does it take you to come to school? Draw the face of the clock; locate the hands at the time you started and again when you reached school. What part of the entire distance around the clock had the minute hand traveled?

7. When the hour hand points between VIII and IX, and the minute hand points to VI, what time is it?

8. Show the position of the hands at 15 minutes past III.

9. If we spend ⅓ of the 24 hours in sleep, how many hours are we awake?

10. If we sleep 10 hours of the day and spend 2 hours at the table, how many hours are left for other duties?

11. What part of this remaining time are we in the schoolroom?

> 12 inches make 1 foot.
> 3 feet make 1 yard.

12. What are the dimensions of the carpenter's steel square or framing square? Draw one on the blackboard.

13. If a carpenter uses the long arm of the square, how many times does he move it in measuring off 18 feet?

14. How many times must a yard measure be moved to measure the same distance?

15. Draw a plan for setting out 20 grape vines in 4 rows so that the vines will be 2 yards apart each way. Let 1 inch represent a yard in your plan.

16. How many vines do you have in each row?

17. Three rows are what part of the total number of vines?

18. How many 2-foot squares are there in a surface 4 feet by 6 feet?

19. How many square feet are there in the same surface?

20. From problem 15, as given above, make 5 original problems.

1. How many hours are there from 9 o'clock am to 4 o'clock pm?
2. Over what part of the circle does the hour hand move from 10 am to 1 pm?
3. If we sleep 8 hours and spend 5 hours in the schoolroom, how many hours of the day remain?
4. How many hours from 4 pm to 9 am the following day?
5. Six hours are what part of the day?
6. Eight hours are what part of the day?
7. One eighth of the day equals _____ hours.
8. Three yards and 5 feet make _____ feet.
9. From a board 4 yards long, 6 feet were cut off. How many feet remained?
10. How many feet in ½ of a yard?
11. How many inches in ½ of a yard?
12. What part of 2 yards are 2 feet?
13. Four yards and 9 feet equal _____.
14. One third of a yard, ⅓ of a foot, and 6 inches equal _____ inches.
15. How tall are you in yards, feet, and inches?

ORAL EXERCISES

$3 \times 5 + 6 - 4 =$ _____ $½$ of $8 \times 3 - 6 =$ _____

$4 \times 5 + 5 - 6 =$ _____ $6 \times 4 - 4 : 5 =$ _____

$3 \times 7 + 4 - 5 =$ _____ $⅓$ of $15 \times 4 - 8 =$ _____

$3 \times 8 - 10 + 2 =$ _____ $18 : 6 \times 5 + 3 =$ _____

$3 \times 6 + 4 + 3 =$ _____ $5 \times 5 - 5 : 5 =$ _____

✓ Use the device in the Sight Exercises section of Review 1 for drill in multiplication and addition, making the combination as high as 25.

✓ What part of 16 is 4? ___   What part of 20 is 5? ___
What part of 21 is 7? ___   What part of 24 is 6? ___
What part of 18 is 6? ___   What part of 21 is 3? ___
What part of 24 is 8? ___   What part of 24 is 4? ___
What part of 21 is 3? ___   What part of 25 is 5? ___

✓ Give the multiplication table of 2's and 3's.

👓 SIGHT EXERCISES

✓ Instructions:

	a	*b*	*c*
	1	2	1
	2	2	2
	2	2	2
	2	2	2
	2	2	1
	2	1	1
	2	3	2
	2	2	1
	2	2	4
	2	1	2
	2	2	3
	2	2	2
	+ 2	+ 2	+ 2

1. Add column *a*, beginning at the bottom.
2. Practice until this can be done accurately in 10 seconds or less.
3. Add column *a*, beginning at the top.
4. Add column *b* in 10 seconds.
5. Add column *c* in 19 seconds.

MEASUREMENTS

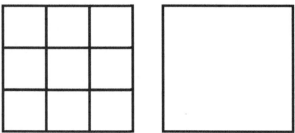

1. There are 3 feet in one yard. How long and how wide is the schoolroom in yards? Measure it with a yard-stick.

2. What are the dimensions, in feet, of a room that is 5 yards long and 4 yards wide?

3. How many square yards of carpet would it take to cover the floor of this room?

4. A man had 24 acres of land which he wished to divide as follows: ½ of it for an orchard, 6 acres for small fruit, ⅛ of it for a garden, and the remainder for house, barn, and lawn. How many acres were devoted to each purpose?

5. Make a drawing showing how this could be done using 1 square inch to represent an acre, and having the drawing 6 inches long. Have the small fruit plot 3 inches long, and the garden plot 2 inches long.

6. What is the area of the top of your desk in square feet?

7. When two pupils occupy the same seat, how many square feet of desk belong to each?

8. What is the area of your arithmetic in square inches?

9. Cut a paper large enough to cover the entire book, allowing one inch to turn in at both the sides and ends. What is the area, in square inches, of the paper?

10. What is the area of each window pane in your schoolroom? What does it cost to replace a broken pane?

11. If all the windows were so arranged as to admit one unbroken sheet of light, what would the area be in square feet?

12. There are 26 states east of the Mississippi River. Fourteen border on the Atlantic Ocean. How many do not?

13. One half as many states border on the Great Lakes as on the Atlantic Ocean. How many border on the Great Lakes?

14. Of these 26 states, 16 are mountainous. How many have no mountains?

15. Find from the map how many states touch the Mississippi River.

16. How many touch the Ohio River? How many touch the Missouri River?

17. Thirteen states border on the Atlantic Ocean, 5 on the Gulf of Mexico, and 3 on the Pacific Ocean. How many states have a seacoast?

18. Paul wrote 14 books of the New Testament, John wrote 5, Peter wrote 2, Luke wrote 2; and the remainder were written by Matthew, Mark, James, and Jude, who wrote 1 each. How many books are there in the New Testament?

19. There are twice as many chapters in the book of Matthew as the number of the books that Paul wrote. How many chapters are there in it?

20. How many more chapters has the book of Matthew than has the book of Revelation?
21. Make 3 original problems in measurements, using objects in the schoolroom.

 DRILL

1. Twenty six less 14 equals _____.
2. One half of 14 is _____.
3. Twenty six less 16 equals _____.
4. Thirteen, 5, and 3 are _____.
5. Twenty three less 5 equals _____.
6. Fourteen, plus 5, plus 2, plus 2, plus four 1's equal _____.
7. Fourteen is ½ of _____.
8. Three 9's are _____.
9. Twenty eight less 22 equals _____.
10. One half of 24 is _____.
11. Six is what part of 24? _____
12. How many yards in 27 feet? _____
13. How many feet and inches in 28 inches? _____
14. How many gallons in 28 quarts? _____
15. How many pints in 13 quarts? _____
16. How many square feet in 1 square yard? _____
17. How many square feet in 3 square yards? _____
18. A nickel is what part of a quarter of a dollar? _____
19. How many dimes in 30 cents? _____
20. A boy having 30 cents spent 2 dimes and a nickel. How much had he left? _____

✓ Add:

3	2	4	5	4	6	6	5
4	3	2	3	4	5	6	5
5	6	5	4	4	5	6	8
+ 5	+ 8	+ 9	+ 4	+ 5	+ 5	+ 6	+ 8

2	3	2	3	2	5	3	4
3	3	4	3	4	5	5	5
3	3	6	4	5	6	6	6
4	6	7	4	5	6	7	7
+ 4	+ 6	+ 7	+ 4	+ 5	+ 6	+ 7	+ 8

✓ 12 – 4 = ___   ✓ 28 – ___ = 18   ✓ 23 + 4 – 9 = ___

16 – 8 = ___   16 – ___ = 12   22 + 5 – 7 = ___

24 – 10 = ___   30 – ___ = 10   24 + 6 –15 = ___

27 – 6 = ___   29 – ___ = 12   20 + 8 – 12 = ___

✓ How many 3's in 24? ___   ✓ 4 is what part of 28? ___

How many 4's in 16? ___   9 is what part of 27? ___

How many 5's in 25? ___   10 is what part of 30? ___

How many 8's in 24? ___   6 is what part of 30? ___

How many 9's in 27? ___   ½ of 30 is what? ___

How many 7's in 28? ___   ⅙ of 30 is what? ___

How many 6's in 24? ___

How many 5's in 30? ___   ✓ Use the device in the Sight Exercises section of Review 1 for drill in multiplication and addition, making the combination as high as 30.

How many 4's in 28? ___

How many 3's in 27? ___

How many 6's in 30? ___

How many 15's in 30? ___

LESSON 10

PROBLEMS IN PHYSIOLOGY

1. Our first, or temporary, set of teeth are 20 in number. Our permanent set has 12 more. How many teeth in our permanent set?

2. How many teeth are there in the lower jaw of the permanent set?

3. How many teeth are there in each half jaw?

4. In each half jaw, there are 3 teeth called grinders, or molars. How many molars in a full set of 32 teeth?

5. How many cutting teeth, or incisors, are there in a full set of 32? How many eye teeth, or canine teeth? How many bicuspids—two pointed?

> How many incisors does the cow have in her upper jaw? What animals do you know that have very large teeth? Do snakes and birds have teeth?

6. If you spend 5 minutes each day brushing your teeth, how many minutes will be thus spent in a week? In how many days would you have spent an hour brushing your teeth?

7. There are 14 bones in the face, 8 in the skull, and 6 in the ears. What is the total number of bones in the head?

8. How many more teeth in the permanent set than there are bones in the head?

9. What part of the entire permanent set of teeth are the wisdom teeth?

10. What part of the entire permanent set are the incisors?

11. The length of the alimentary canal in a human being is about 5 times his height. How long is it in a person 5 feet tall?

12. How long is it in a person 6 feet tall?

13. If the mouth, esophagus, and stomach taken together are 2 feet long, and the large intestine 5 feet long, how long is the small intestine in a person 5 feet tall?

> The word *esophagus* is from the Greek *oisein* - to carry, and *phageīn*-to eat; to carry what is eaten. The passage for food.

14. The alimentary canal is usually longer in the vegetable-eaters and shorter in the flesh-eaters. In the cat and dog family, it is about 3 times the length of the body. How long is it in a lion 7 feet long?

15. Complete the following multiplication table.

3 x 1 = 3		3 x 5 = ___		3 x 9 = ___	
3 x 2 = 6		3 x 6 = ___		3 x 10 = ___	
3 x 3 = ___		3 x 7 = ___		3 x 11 = ___	
3 x 4 = ___		3 x 8 = ___		3 x 12 = ___	

16. Make one example each in addition, subtraction, and fractions from physiology.

DRILL

1. Twenty teeth plus 12 teeth are _____ teeth.
2. One fourth of 32 equals _____.
3. Three 4's are _____.
4. Five times 7 are _____.
5. Four and one fourth inches minus 2 inches are _____ inches.
6. Fourteen, 8, and 6 are _____.
7. Fourteen, 5, 8, 3, and 2 are _____.
8. Five times 6 = _____
9. Two and two-thirds feet equal _____ inches?
10. Thirty three feet are _____ yards?
11. One third of a yard, 1 foot, and 10 inches equal _____ inches?
12. How many posts, set 2 yards apart, are used in making a fence 34 yards long?
13. How many yards around a square garden, one side of which measures 8 yards?
14. How many boards will be needed to fence this garden, each board being 12 feet long, and the fence being 5 boards high?
15. How many strawberry vines can be set in this garden, if the vines are placed two feet apart?
16. Eight yards are _____ feet.
17. Five 2's are _____, and four 10's are _____.

✓ ⅕ of 30 = ___	⅐ of 14 = ___	⅕ of 20 = ___
¼ of 32 = ___	⅕ of 35 = ___	½ of 26 = ___
⅓ of 27 = ___	⅙ of 30 = ___	⅐ of 35 = ___
⅙ of 24 = ___	⅛ of 32 = ___	⅔ of 30 = ___
¾ of 12 = ___	⅖ of 20 = ___	⅗ of 25 = ___

Instructions:

1. Add column *a* in 10 seconds.
2. Practice on each column until you can add it in 10 seconds, beginning either at the top or bottom. Be accurate.

a	b	c	d	e
1	3	2	1	1
2	2	2	2	2
2	3	2	3	2
2	2	2	1	1
2	2	2	2	1
2	2	2	3	3
2	3	3	2	1
2	2	3	3	1
2	1	3	1	3
2	3	3	2	4
2	2	3	3	4
2	3	3	4	4
2	2	3	5	4
+ 3	+ 2	+ 3	+ 4	+ 4

Instructions:

Place the following device on the blackboard to be a drill in addition, subtraction, multiplication, and division. The central figure may be changed from time to time.

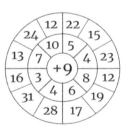

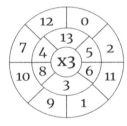

LESSON 11

PROBLEMS IN WEIGHT

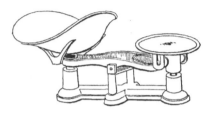

1. Sixteen ounces make 1 pound. How many ounces in 2 pounds?

2. A pint of water weighs 1 pound. How much will 1 gallon weigh?

3. Mercury, or quicksilver, is about 13½ times as heavy as water. How much would a quart of mercury weigh?

> Gold is about 19 times as heavy as water. Milk and blood are about as heavy as water.

4. If a 10-gallon can is half full of milk, how much will the milk weigh?

5. The average adult person has about a gallon and a half of blood. How many pounds of blood has he?

6. If ¼ of the blood is distributed to the heart and large blood vessels, ¼ to the liver, and ¼ to the muscles, how many pounds remain for the other organs?

7. If I pay 12 cents for 32 ounces of sugar, what is the price per pound?

8. How many pounds of sugar can I buy for 36 cents at the same price per pound?

9. In preparing grains for the table, 3 quarts of water added to 1 quart of rice makes 3 quarts of cooked rice. If from each quart of cooked rice 3 orders are served, how many orders does each quart of dry rice make?

10. If each order sells for 2 cents, how much is received for 1 quart of dry rice?

11. How many quarts of dry rice does it take for 36 orders?

12. In preparing corn meal, 4 quarts of water are added to 1 quart of dry corn meal, which makes, when cooked, 4 quarts of grain. If 3 orders are served from each quart, how many orders does each quart of dry corn meal make?

13. How many quarts of dry corn meal does it take to prepare orders for 36 people?

14. At 2 cents an order, what is received for each quart of corn meal?

15. Three quarts of water added to 2 quarts of rolled oats make 3 quarts of the cooked grain. If each quart of cooked oats makes 3 orders, how many orders may be obtained from 4 quarts of dry rolled oats?

16. How many quarts of dry rolled oats must be used in preparing 36 orders?

17. What is received for each quart of dry rolled oats, when cooked and served at 2 cents an order?

18. One quart of water added to 1 quart of crystal wheat makes one quart of cooked crystal wheat, from which 4 orders are served. How many orders can be obtained from 6 quarts of dry crystal wheat?

19. Five quarts of water added to 2 quarts of gluten make 5 quarts of cooked gluten. If 4 orders are obtained

from each quart of cooked gluten, how many order does 1 quart of dry gluten make?

20. How many quarts of dry gluten must be used in preparing 50 orders?

 DRILL

1. How many ounces in 2 pounds?
2. One and one-half pounds less 4 ounces equal _____ ounces.
3. How many pounds does 2 gallons of water weigh?
4. About how much does 5 gallons of milk weigh?
5. About how much does the blood in your body weigh?
6. Thirty two divided by 16 = _____
7. How many quarts of beans can I buy for 40 cents at 5 cents a quart?
8. How many dimes in 40 cents?
9. A quarter of a dollar, a dime, and 3 cents equal _____ cents.
10. Forty cents less a quarter of a dollar equal _____ cents.
11. What part of a pound are 4 ounces?
12. Two and one-half pounds equal _____ ounces.

> 10 cents make 1 dime.
> 10 dimes make 1 dollar.

13. Three quarters, 2 dimes, and 1 nickel make _____ dollars.

✓ Add:

3	4	3	5	4	3	5	5
4	4	5	7	5	6	5	5
6	8	7	8	7	8	5	6
+ 10	+ 12	+ 14	+ 12	+ 15	+ 14	+ 16	+ 18

6	3	4	3	5	4	3	4
5	5	8	6	4	5	7	2
4	4	5	4	2	6	2	1
10	7	3	8	7	8	1	8
+ 12	+ 16	+ 18	+ 20	+ 21	+ 22	+ 25	+ 24

ORAL EXERCISES

✓ 24 + 6 : 5 - 3 = ___   5 x 7 + 1 : 9 - 4 = ___

32 + 4 : 6 x 4 = ___   5 x 5 + 5 : 5 - 5 = ___

½ of 16 x 15 - 7 = ___   ½ of 30 : 5 x 6 + 2 = ___

35 : 7 x 4 + 6 = ___   ⅔ of 21 + 2 : 4 - 3 = ___

4 x 9 + 4 : 10 = ___   ⅗ of 30 : 2 + 1 - 5 = ___

✓ Use the devices in the Sight Exercises section of Review 1 and the Drill section of Lesson 10.

✓ Repeat the multiplication table of 2's, 3's, 4's, and 5's.

✓ 4 x 5 = ___	4 x 3 = ___	3 x 8 = ___
2 x 10 = ___	5 x 7 = ___	4 x 9 = ___
5 x 6 = ___	5 x 4 = ___	5 x 2 = ___
4 x 8 = ___	4 x 6 = ___	2 x 12 = ___
3 x 7 = ___	4 x 3 = ___	4 x 12 = ___
2 x 9 = ___	2 x 11 = ___	5 x 11 = ___

✓ A NUMBER EXERCISE

The effectiveness of a drill is often increased if so conducted that the pupil forgets for the time being that he is doing ordinary arithmetical work.

For this game, prepare cards of thin cardboard. On each one, write a simple question or statement in arithmetic, as 6x7 =, 4x8 =, 16:4 =, 4 pecks make _____, etc., etc. Grade the questions to suit the advancement of the pupils.

$5 \times 8 =$ $16 - 4 =$

$\frac{3}{4}$ of $20 =$ How many inches in 2 feet?

$21 - 7 =$ How many pecks in $2\frac{1}{2}$ bushels?

$25 + 12 =$ 2 feet-4 in =

Specimen Cards

Place the cards face down in a promiscuous heap in the center of a table, and allow each child to draw one, which he places behind him without reading.

The exercise begins by asking one pupil to read and answer the question on the card which he holds. Having done so, the card is laid on the table before him, and another card drawn from the heap and held at his back.

The pupils recite in turn. If a mistake is made, nothing is said at the time, but the one who notices it awaits his turn, and then corrects the mistakes and takes the card from the mistaken pupil, answering also his own card. If a pupil remembers more than one error in answers, he is free to correct them all. The work is wholly mental and holds the attention of the pupils. In the end, the one who holds the largest number of cards has the best standing.

DRY MEASURE

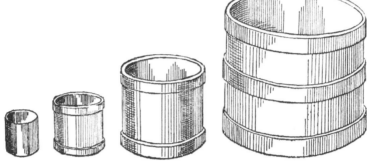

1. These measures are used for measuring grains, seeds, nuts, potatoes, and other dry articles. Measure and find how many times the pint measure must be filled in order to fill the quart measure.

2. How many times will the peck measure fill the quart measure?

3. Having the bushel measure full of apples, how many times can you fill the peck measure?

4. A man picked apples into a half-bushel basket and filled it 16 times, how many bushels had he?

5. He sold these apples at 50 cents a bushel, what did he receive for them ?

6. When potatoes sell for 40 cents a bushel, what does a peck cost?

7. James filled the peck measure with potatoes 16 times while his brother filled the bushel measure 3 times. Which boy was the more rapid worker? What was the difference in pecks? in bushels?

8. These boys were paid 2 cents a bushel for picking up potatoes. When James had earned 24 cents, how much had his brother earned?

9. In sorting apples, Thomas filled the peck measure with decayed apples every time he filled the bushel measure with sound apples. When he had 16 bushels of sound apples, how many decayed apples had he found?

10. Reuben was told to give each cow 6 quarts of bran for her supper. How many cows can he feed from 3 peck measures full of bran?

11. A bushel of oats weighs 32 pounds. How many pounds does 1 peck weigh?

12. How much does 1 quart of oats weigh?

13. A bushel of barley weighs 48 pounds. How many times heavier is barley than oats?

14. One bushel of beans weighs 60 pound. How much does 1 peck weigh?

> 2 pints make 1 quart.
> 8 quarts make 1 peck.
> 4 pecks make 1 bushel.

 BIBLE MEASURES

1 cor, measure, or homer	=	7½ bushels
1 ephah	=	3 pecks

1. How many bushels of wheat was Ezra, the scribe, allowed by the Persian king to require for the house of God? Ezra 7:22.

2. Express in quarts and the fractions of a quart the amount of flour required as a sin offering of the poor in Leviticus 5:11.

3. Gideon went in, and made ready a kid, and unleavened cakes of an ephah of flour (Judges 6:19). How many quarts did he use in making the cakes?

4. What part of a bushel of barley did Ruth glean in one day? Ruth 2:17.

5. What is the weight in pounds of the barley she gleaned in one day?

6. If Ruth had gleaned oats instead of barley, what would have been the difference in the weight of the grain she carried home at night?

7. In days of prosperity, the land of Palestine yielded a hundredfold. If 1 homer of seed was planted, what would be the number of bushels reaped?

8. When Israel disobeyed God, the seed of a homer yielded but an ephah. Isaiah 5:11. In this case, how many pecks did a homer yield? What part is this of the seed sown?

DRILL

1. Eight pints make _____ quarts.
2. Nine pints make _____ quarts.
3. Eight quarts of oats fill the pint measure _____ times.
4. Sixteen quarts of strawberries fill the peck measure _____ times and the pint measure _____ times.
5. Sixteen pecks of potatoes fill the bushel measure _____ times.
6. Twelve pecks of potatoes, at 40 cents a bushel, cost _____ cents.
7. These 12 pecks of potatoes, at 40 cents a bushel, can be paid for with _____ dimes.
8. There are _____ quarts in 1½ pecks.
9. How many quarts in 1 bushel?
10. What part of a bushel is 1 quart ?
11. One fourth of a bushel is called a _____.
12. One half of a quart is called a _____.
13. One eighth of a peck is called a _____.
14. What is ¹⁄₁₆ of 1 peck called?
15. What is ¹⁄₃₂ of 1 bushel called?
16. One half of a bushel equals _____ pecks.

SIGHT EXERCISES

✓ Add:

3	5	8	12	13	9	11
7	4	2	8	7	1	9
6	1	6	6	3	8	14
4	7	4	4	7	2	6
+ 10	+ 3	+ 3	+ 3	+ 8	+ 15	+ 3

✓ Subtract:

32	42	16	29	18	25
– 16	– 12	– 11	– 14	– 12	– 17

✓ 32 : 2 = ___   ✓ ⅓ of 30 = ___

32 : 4 = ___   ½ of 20 = ___

32 : 8 = ___   ⅔ of 30 = ___

32 : 16 = ___   ⅛ of 32 = ___

16 : 4 = ___   ⅕ of 15 = ___

25 : 5 = ___   ⅖ of 15 = ___

30 : 2 = ___   ¼ of 32 = ___

35 : 7 = ___   ½ of 32 = ___

60 : 12 = ___   ⅕ of 25 x 6 = ___

55 : 11 = ___   2/7 of 14 x 5 = ___

44 : 4 = ___   ⅙ of 36 x 2 = ___

21 : 3 = ___   ⅛ of 32 x 4 = ___

✓ Repeat the multiplication table of 5's and 6's.

> **NOTE**
> Make use of card game previously given.

PROBLEMS IN PAPERING

1. How many square yards of carpet would it take to carpet a room 15 feet long and 12 feet wide?

2. If the room is 10 feet high, how many square yards of wallpaper would it take to paper one side?

3. How many square yards on the opposite side, deducting six square yards for a door and window?

4. How many square yards in both ends, deducting 4 square yards for 2 windows?

5. How many square yards of wallpaper would it take to paper the four walls of the above room?

6. How many square yards would it take for the ceiling?

> Wallpaper may be bought in single or double rolls. The ordinary paper is from 18 to 22 inches wide. Single rolls are 8 yards long, and double rolls are 16 yards long.

7. How many feet long is a single roll? How many square feet in a single roll which is 18 inches, or 1½ feet, in width?

8. How many square yards are there in such a roll?

9. How many such rolls would it take to paper the four walls of the house just mentioned?

10. How many rolls would it take for the ceiling?

11. How much would 16 rolls cost at one fourth of a dollar a roll?

12. How many square feet of blackboard in the schoolroom?

> Estimate it with your eye, then measure. What is the difference in results?

13. Make an original problem in which you estimate the cost of papering the schoolroom.

REVIEW 2

1. How many school days are there in 4 weeks? In 8 weeks?
2. How many hours are you in the schoolroom each day? How many each week?
3. How many hours from 9 o'clock am Monday till 1 pm the next Tuesday?
4. How many hours from 6 pm till 11 am the following day?
5. In order that the stomach may have sufficient time to properly digest food, and also have time for rest, physicians advise a seven-hour interval between meals. If breakfast is eaten at 6:30 am, at what hour should dinner be eaten?
6. What would be the supper hour?
7. As three hours should be allowed for the stomach to do its work before sleep, what would be the retiring hour? Would the rising hour be in time for breakfast at 6:30 the next morning?
8. How many times is the foot rule moved in measuring a table 48 inches long?
9. What is the length of the same table in yards?
10. What part of 1 square yard is 1 square foot? Make a drawing to illustrate this.
11. How many square feet in ⅔ of a square yard? Fold a square of paper to show this.
12. A window is 6 feet long and 3 feet wide. What will a shade for this window cost at 3 cents per square foot and 10 cents for a roller?
13. Three boys picked apples. The first picked 2 bushels and 8 quarts; the second picked 1 bushel

and 4 pecks; and the third picked 3 bushels; how many bushels did the three boys pick?

14. How many times can a lamp which holds 1 quart of oil be filled from a 5-gallon can?

15. When this lampful of oil burns 2 evenings, how often does the 5-gallon can have to be filled?

16. When oil sells for 12 cents a gallon, what does it cost to fill the lamp once?

17. What does the 5-gallon can of kerosene cost?

18. The tank on gasoline stove holds 1 gallon. Gasoline sells at 10 cents a gallon. The tankful of gasoline cooks 5 meals on an average. The stove is used twice each day. Make 2 original problems from these items.

19. After Ruth gleaned in the barley harvest one week (Ruth 2:17), how many bushels of barley had she?

20. How many pounds of bread in four 12-ounce loaves?

21. How much will 32 ounces of bread cost at 5 cents a pound?

22. How much must I pay for 1 peck of oats at 40 cents a bushel?

23. How many pounds of oats in half a bushel?

24. How many sacks would it take to hold 48 bushels of oats if each sack holds 3 bushels?

25. At 8 cents a pound, how many pounds of rice may be bought for 40 cents?

26. Four fifths of rice being starch, how many pounds of starch would there be in 40 cents' worth?

27. How many pounds of starch in 50 pounds of rice?

28. Repeat the multiplication table of 4's and 5's.

29. One bushel of oats weighs 32 pounds; 1 bushel of barley weighs 56 pounds; 1 bushel of wheat weighs 60 pounds. Make 2 problems from these items.

 DRILL

1. Five times 4 are _____.
2. Twelve and 4 are _____.
3. Fifteen feet equal _____ yards.
4. Three times 4 square yards equals _____ square yards.
5. Fifteen, 10, and 20 are _____.
6. One and a half times 24 = _____
7. Twenty three states and 26 states are _____ states.
8. One half of 26 states equals _____ states.
9. Three times 9 less 14 equals _____.
10. Twenty seven books and 12 books equal _____ books.
11. Thirty two teeth and 8 teeth are _____ teeth.
12. How many pounds do 6 gallons of water weigh?
13. Four times 12 ounces equal _____ pounds.
14. Thirty two divided by 16 multiplied by 5 = _____
15. One half of 32 pounds equals _____ equals.
16. Three is contained in 48 _____ times.
17. 40 divided by 8 = _____
18. Four fifths of 5 = _____
19. Four fifths of 50 = _____
20. One half of a dollar less a nickel and a dime equals _____ cents.

✓ Use devices already given for drill in multiplication, division, and subtraction.

✓ **Instructions:**

1. If possible, add column *a* in 15 seconds, beginning at the bottom.
2. Add column *a* in 15 seconds, beginning at the top.
3. In the same way, add the other columns.

Four 10's are _____

a	b	c	d
1	2	1	1
2	2	2	2
2	2	1	3
2	1	3	2
2	2	1	1
2	1	1	2
2	3	2	3
2	3	1	4
2	3	1	5
2	3	3	2
2	3	1	1
2	3	1	3
2	3	2	1
2	3	1	2
2	3	4	2
2	3	1	2
2	3	2	1
2	3	4	2
3	3	4	5
3	3	4	5
+ 3	+ 3	+ 4	+ 5

✓ 1 whole is _____ fourths.

1 whole is _____ eighths.

1 whole is _____ thirds.

1 whole is _____ sixths.

1 fourth is _____ eighths.

2 fourths are _____ eighths.

3 fourths are _____ eighths.

4 fourths are _____ eighths.

1 third is _____ sixths.

2 thirds are _____ sixths.

3 thirds are _____ sixths.

6 sixths are _____ eighths.

3 thirds are _____ eighths.

✓ MEASURING ASSIGNMENT

Measurement	Estimate	True Measure	Error	Percentage of Error
Height of ceiling				
Length of teacher's desk				
Diagonal distance across room				
Width of pupil's desk				
Circumference of stovepipe				
Diameter of water pail				
Number of square yards in blackboard				
Distance around the head				
Width of door				
Length of window				

Let the teacher copy the above diagram on the blackboard. Have the pupil estimate the measurements of the different objects indicated, the teacher placing their estimates in the proper column. Afterward, test the estimates with a yard stick, marking the error. Advanced pupils may fill out the last column. Similar drills should be frequently given.

LESSON 14 — AREA OF TRIANGLE

1. Cut out a piece of paper 4 inches square. How many square inches does it contain?

2. Fold the paper so that the upper right-hand corner meets the lower left-hand corner. Crease, and cut on the crease. Each of the figures thus made is a right triangle. Describe a right triangle. How many right triangles in a square?

3. How many square inches do each of these contain? What is the length of the 2 short sides?

4. Fold the triangle so that the vertex of the upper angle will meet the vertex of the right angle. Crease and open. Mark the upper part A, and the lower part B. Tear off A, and place it next to B so that A's long side will be next to B's side of the same length, A being upside down. How do you find the area of this figure?

5. The width of this figure is what part of the height of the triangle?

6. Make a rule for computing the area of right triangles.

7. How many square inches in a right triangle whose base is 12 inches, and whose altitude, or height, is 8 inches?

8. Draw a horizontal line 16 inches long, dividing it in the midde. Mark one end A, the other end B, and the middle D. From D, draw a line perpendicular to AB six inches long. Mark the top of this line C. Draw lines from C to A and from C to B. How many right triangles are there? What is the area of each triangle?

9. Let the figure ABC represent the gable end of a house, 1 inch representing a foot. How many square feet of lumber would it take to cover this end?

10. Compute the area in the gable end of a house that is 20 feet wide, the peak of the roof being 8 feet higher than the eaves of the house.

11. James had 50 cents. He bought 5 pounds of oatmeal at 4 cents a pound, and spent the remainder of his money for sugar. How much did he spend for sugar?

12. The sugar being 6 cents a pound, how many pounds did he buy?

13. How much would 1 peck of raspberries cost at 7 cents a quart?

14. How many times can a 2-quart measure be filled from 3 pecks of berries?

15. If I have a quarter of a dollar, what other 2 pieces of money must I have in order to pay a bill of 55 cents?

16. What would 7 pounds of prunes cost at 8 cents a pound?

17. What 3 pieces of money may be used in paying for them?

18. _____ pints make 1 quart.

   _____ quarts make 1 gallon.

   _____ quarts make 1 peck.

   _____ pecks make 1 bushel.

   32 quarts make 1 _____.

   1 quart is what part of 1 gallon?

   2 quarts are what part of 1 gallon?

   2 quarts are what part of 1 peck?

19. Which of these measures are used for grains? Which for liquids?

DRILL

1. What is the square of 4?

2. One half of the square of 4= _____   ½ of 16 = ___

3. ½ of 6 X 8 = _____
   ½ of 8 X 10 = _____

4. A right triangle has _____ sides and _____ right angles.

5. What is the area of a right triangle whose base is 8 inches and whose height is 4 inches?

6. Give the dimensions of a rectangle having the same area.

7. The area of a right triangle is 32 square inches. What is the length of one side of the square of which this triangle is one half?

8. A quarter of a dollar and 3 dimes equal _____ cents.

9. With half a dollar, how much change would I receive after paying for 7 quarts of berries at 6 cents a quart?

10. A quarter, a dime, a nickel, and 3 cents equal _____ cents.

11. How many square feet in 3 square yards?

12. How many square feet in 6 square yards?

13. How many inches in 4½ feet?

✓ **Instructions:**

1. Add each of the columns in 15 seconds, beginning at the bottom and then at the top.

   Three 4's and three 5's = _____

   Five 7's less three 5's = _____

a	b	c	d
1	1	5	1
3	2	5	2
3	4	5	3
3	3	5	4
3	1	5	5
3	2	5	1
3	1	2	2
3	1	1	3
3	2	2	2
3	3	3	1
3	1	1	2
3	4	1	1
3	4	3	3
3	4	3	6
3	4	3	6
3	4	3	6
+3	+4	+3	+6

ORAL EXERCISES

✓ 7 x 8 – 6 : 2 = _____

4 x 12 + 2 – 25 = _____

½ of 18 x 6 –4 = _____

6 x 9 – 14 : 2 = _____

4^2 –1 : 3 x 5 = _____

5 x 10 – 10 : 10 = _____

6 x 10 – 10 : 2 = _____

6 x 9 + 5 – 30 = _____

2 x 30 + 10 – 15 = _____

6 x 11 –20 : 7 = _____

✓ ½ = how many 4ths? _____

¾ = how many 12ths? _____

½ = how many 12ths? _____

¼ = how many 12ths? _____

¾ = how many 12ths? _____

⁵⁄₅ = how many 12ths? _____

How many 5ths = 1 whole? _____

How many 10ths = 1 whole? _____

⅕ = how many 10ths? _____

³⁄₅ = how many 10ths? _____

²⁄₅ = how many 10ths? _____

½ of ½ = how many 12ths? _____

PROBLEMS IN BREATHING & TIME

1. There are 60 seconds in a minute. How many seconds in ½ of a minute?

2. If you breathe once every three seconds, how many times will you breathe in ½ of a minute? In 1 minute?

3. Have someone test the number of times you breathe in 1 minute. Exercise vigorously for a minute, and then make the test. How many more times did you breathe after exercising than before? Why is this difference?

4. About 6 quarts of air pass into and out of the lungs every minute. How many quarts pass in 10 minutes?

5. In 5 quarts of air, there is 1 quart of oxygen. How many quarts of oxygen are inhaled every 10 minutes?

6. The oxygen in the air is used by the lungs to purify the blood. Only about ¼ of the oxygen inhaled at each breath is utilized. About how many quarts of oxygen pass into the blood from the lungs every 10 minutes?

7. How many gallons of nitrogen in 60 quarts of air if ⅘ of the air is nitrogen? Of what use is the nitrogen in the air?

8. Air that has been exhaled by the lungs has lost about ¹⁄₂₀ of its bulk of oxygen and gained an equal amount of carbonic acid gas, which is a deadly poison. How many quarts of this gas in air that has been breathed by one person 10 minutes?

9. In the lungs of the average adult, there are always about 3 pints of air that cannot be expelled. If the

person is able to inhale at a deep breath 1 gallon of air, how many pints are there then in the lungs?

10. There are sixty minutes in an hour. How many minutes in ¼ of an hour?

11. How many minutes are marked off by the minute hand of a clock in passing from XII to I?

12. How many minutes are marked off by the minute hand in passing from VIII to XII?

13. How many minutes does it take the minute hand to complete the circle?

14. How long does it take the hour hand to travel as far?

15. The minute hand moves how many times faster than the hour hand?

16. How long does it take you to go home from school? What time is it when you reach home?

17. The carrier pigeon's average rate of flight is about 30 miles an hour. How far can it fly in 20 minutes?

 DRILL

1. Thirty is what part of 60?
2. How many 3's in 30?
3. Fifteen seconds equal what part of a minute?
4. About how many times do you breathe in 1 minute?
5. If your heart beats 1½ times every second, how many times does it beat in ½ a minute?
6. How long does it take you to walk a mile?
7. If a man walks a mile in 15 minutes, how far does he walk in 1 hour?
8. If a horse travels 8 miles an hour, how far does he go from 6 am to 1 pm?
9. At the rate of 60 miles an hour, how far does a passenger train run in 2 minutes?
10. At the rate of 30 miles an hour, how far does a freight train run in 1 minute?
11. If a boy earns 5 cents an hour, how much does he earn in ⅓ of 24 hours?
12. If a man works every working day in the week for 4 weeks at 1½ dollars per day, how much does he earn?
13. How many school days are there in 9 weeks? How many weeks in a school year of 9 months, counting 4 weeks to the month?

✓ Count by 3's from 3 to 60.
Count by 3's from 1 to 58.
Count by 3's from 2 to 59.
Count by 4's from 4 to 60.
Count by 4's from 1 to 57.
Count by 4's from 3 to 57.
Count by 5's from 5 to 60.
Count by 5's from 1 to 56.
Count by 5's from 2 to 57.
Count by 5's from 3 to 58.

✓ **MULTIPLICATION & DIVISION**

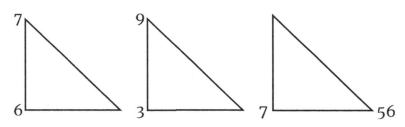

In multiplying, say 7 x 6 or 6 x 7 = 42, and place the answer at the third angle. Likewise, 3 x 9 = _____. The teacher uses a pointer and changes the figures rapidly. In division, begin with the larger number, and reverse the operation.

✓ ½ = how many fourths?   _____

½ = how many eighths?   _____

¼ = how many eighths?   _____

¾ = how many twelfths?   _____

⅕ = how many tenths?   _____

⅔ = how many ninths?   _____

⅗ = how many tenths?   _____

⅘ = how many twentieths?   _____

TEMPERATURE PROBLEMS

1. Examine a thermometer (heat-measure: from the Greek *therme*—heat, and *metron*, measure), and then draw one, marking the degrees on its face.

2. What number does the column of mercury reach if the thermometer is on ice or in ice-cold water?

3. When the mercury column reaches 42° above zero, how many degrees is it above the freezing point?

4. 24° above the freezing point is temperate. How many degrees above zero is that?

5. On a summer day, the thermometer registered 98°. How many degrees is that above temperate?

6. When the mercury stands at 16° above zero, how many degrees below the freezing point is it?

7. What is the difference in the temperature of the room on a level with your head and the temperature on the floor? How many degrees above freezing is it in each place?

8. At sunrise one morning, it was 4° below zero, and at noon 35° above zero. How many degrees had the mercury risen?

9. At sunset, the temperature was 45° above zero. The next morning, it was 12° above zero. How many degrees had the temperature fallen during the night?

10. Sea water freezes only when the temperature is reduced to 26½°. How much lower is that temperature than the freezing point for fresh water?

11. At great depths, the temperature of the ocean is uniform at about 33½°. How many degrees is this above the freezing point of ocean water?

12. The temperature on the warmest day of the month was 88°, and on the coldest day, the thermometer registered 32°, what was the average temperature of the month?

> **NOTE**
> Find the difference between the warmest and the coolest days. Divide this difference by 2, and add the result to the 32°.

The following table shows the average rainfall in various cities.

Cities	Average Rainfall	Cities	Average Rainfall
New Orleans	51 inches	Toronto	35 inches
New York	45 inches	San Fransisco	23 inches
Milwaukee	30 inches	El Paso	11 inches

13. Express in feet and fractions of feet the amount of rainfall for each of these places.

14. Find the difference in the amount of rainfall at New Orleans and each of the other cities.

15. Make two problems based upon the items given in the above table.

16. Note the difference in temperature between morning, noon, and night, and make three original problems.

DRILL

In the following temperature problems, the sign + means above zero, and the sign - below zero.

1. What is the difference in temperature between—

    56°+ and 32°+?

    30°+ and 12°-?

    70°+ and 32+?

    25°+ and 15°-?

    40°+ and 16°-?

2. How many degrees above freezing is 60°+?

3. How many degrees below freezing is 10°+?

4. 20°- is how many degrees below freezing?

5. How many degrees above zero is temperate?

6. 20° below temperate heat is how many degrees above freezing?

7. 24 + 24 + 4 + 1 + 1 = _____

8. 32 + 32 = _____

9. 32 - 14 = _____

10. ½ of 60 = _____

11. 60 - 54 = _____

12. 14 + 8 + 6 -30 = _____

13. ⅒ of 70 = _____

14. One half of a dollar and 2 dimes are _____ cents.

15. Seventy cents less a quarter of a dollar equals _____ cents.

✓ Add:

3	2	3	1	2	3	4	3
4	4	5	5	4	9	5	6
5	6	7	6	9	8	5	7
6	7	8	9	9	10	11	12
+ 6	+ 7	+ 8	+ 9	+ 9	+10	+ 11	+12

2	3	4	2	3	7	4	3
5	6	5	8	5	8	6	7
6	7	8	9	9	10	11	12
6	7	8	9	9	10	11	12
+ 6	+ 7	+ 8	+ 9	+ 10	+10	+ 11	+12

NOTE

In adding such columns as the last, add thus: 12, 24, 36, 43, 46.

 ORAL EXERCISES

✓ 24 + 12 : 6 – 6   = _____

3 x 9 + 3 + 30   = _____

5 x 8 + 2 : 7   = _____

2 x 25 + 6 :7   = _____

5 x 9 + 5 : 5   = _____

7 x 9 – 3 : 3 +5   = _____

20 + 16 –4 : 8   = _____

½ of 64 + 4 : 9   = _____

9 x 8 – 2 : 2 + 1   = _____

½ of 70 + 1 –16 : 4   = _____

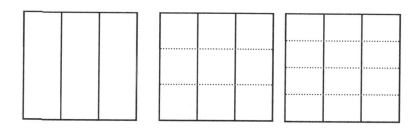

✓ 1 whole = how many 3rds? _____

1 whole = how many 12ths? _____

1 whole = how many 9ths? _____

⅓ = how many 12ths? _____

⅓ = how many 9ths? _____

⅔ = how many 12ths? _____

⅔ = how many 9ths? _____

³⁄₃ = how many 12ths? _____

³⁄₃ = how many 9ths? _____

¼ of ⅓ = _____

⅓ of ⅓ = _____

½ of ⁸⁄₁₂ = _____

½ of ⁶⁄₉ = _____

³⁄₉ = how many 12ths? _____

TIME & WEIGHT PROBLEMS

> Thirty days hath September.
> April, June, and November.
> All the rest have thirty one.
> Except February alone.

Februay has 29 days in a leap year, and 28 days in all other years.

1. In a leap year, how many days are there in January and February together?
2. What day of what month is your birthday? How many days from your birthday to the last day of the month in which it comes?
3. How many days from your birthday to the 25th day of the next month?
4. The first day of August 1900 was Wednesday. The fourth Wednesday in August was what day of the month?
5. What day of the week was the 24th of August 1900?
6. If school began Monday, September 3, upon what day of the month would the first school month end?
7. How many weeks and days are there in January? in February? in April?
8. How many days from September 1 to October 15?
9. How many days from October 15 to December 5?

10. Wheat, clover seed, peas, beans and potatoes all weigh 60 pounds to the bushel. How many pounds do 3 pecks of potatoes weigh? What are they worth at 36 cents a bushel?

11. As only ¼ of potatoes is solid food material, and the balance water, how many pounds of solid food in 1 bushel of potatoes? How many pounds of water?

12. At 64 cents a bushel, what is the price of wheat per peck?

13. If in every bushel of ordinary wheat there are 9 pounds of waste, and the remainder is solid food material, how much solid food is there in 1 bushel of wheat?

14. How much do 1⅓ bushels of peas or beans weigh?

15. As there is about the same amount of waste in peas and beans as in wheat, how much waste is there in 1⅓ bushels of peas or beans?

16. How much solid food material is there?

17. At 5 cents a quart, how much do 1½ pecks of beans cost?

18. If I pay for them with a half dollar and a quarter, how much change do I receive?

 DRILL

1. Thirty one days and 29 days are _____ days.
2. Thirty one days less 18 days equal _____ days.
3. One plus four 7's = _____
4. Three plus three 7's plus 4 = _____
5. Thirty one days equal _____ weeks and _____ days over.
6. Thirty days plus 15 days equal _____ days.
7. 31 days – 15 days + 30 days + 5 days = _____
8. ¼ of 60 x 3 = _____
9. 3 times 25 cents = _____
10. 60 – ¼ of 60 = _____
11. 64 : 32 = _____
    60 – 9 = _____
12. 60 x 1⅓ = _____
    80 – (1⅓ x 9) = _____
13. 8 + 4 – 5 cents = _____
14. 50 cents + 25 cents – 60 cents = _____
15. A quarter of a dollar and 4 dimes equal _____ cents.
16. A half dollar, three dimes, and 3 cents equal _____ cents.
17. Seventy five cents less a quarter and a dime equal _____ cents.
18. Eighty cents less 7 nickels equal _____ cents.
19. Three quarters and a nickel equal _____ cents.

Instructions:

1. Add column *a* in 15 seconds, beginning first at the bottom and then at the top.
2. Do the same with each of the other columns.

a	b	c	d
1	3	3	5
3	4	4	5
4	5	5	5
3	3	6	6
4	4	3	6
3	5	4	6
4	3	5	4
4	4	6	4
4	5	6	6
4	5	6	6
4	5	6	6
4	5	6	6
4	6	6	6
+ 4	+ 6	+ 6	+ 6

ORAL EXERCISES

$6 \times 8 + 12 - 30$ = _____

$7 \times 8 + 4 : 3$ = _____

$8 \times 8 + 6 - 35$ = _____

$6 \times 12 + 8 : 4$ = _____

$5 \times 12 + 2 - 32$ = _____

$8 \times 9 + 3 : 25$ = _____

$9 \times 9 - 1 : 4 + 5 - 15$ = _____

$8^2 - 4 : 2 + 6 : 6$ = _____

$\frac{1}{2}$ of $18 \times 6 + 2 : 7$ = _____

$4 \times 8 \times 2 + 6 : 35$ = _____

$\frac{2}{3}$ of $9 \times 8 + 2 : 10$ = _____

$7 \times 9 + 7 - 40 : 6$ = _____

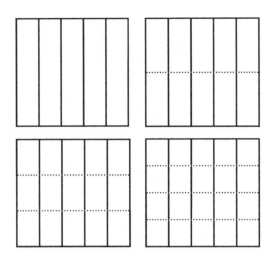

✓ 1 whole = how many fifths? _____
  ⅕ = how many tenths? _____
  ⅖ = how many tenths? _____
  ⅗ = how many tenths? _____
  ⅘ = how many tenths? _____
  ⁵⁄₅ = how many tenths? _____
  ⁵⁄₅ = how many fifteenths? _____
  ⅕ = how many fifteenths? _____
  ⅖ = how many fifteenths? _____
  ⅗ = how many fifteenths? _____
  ⅘ = how many fifteenths? _____
  ⅕ = how many twentieths? _____
  ⅖ = how many twentieths?
  ⅗ = how many twentieths?
  ⅘ = how many twentieths?

✓ $\frac{3}{15}$ = __/20
  $\frac{4}{10}$ = __/20
  $\frac{9}{15}$ = __/20
  $\frac{3}{15}$ = __/10
  $\frac{4}{10}$ = __/15
  $\frac{12}{15}$ = __/20

NOTE
Make use of the card exercise (page 58) as a drill in fractions.

1. With paper and ruler, show how to find the area of a right triangle.

2. The following is a diagram of a piece of land drawn on the scale of ¼ inch to the yard.

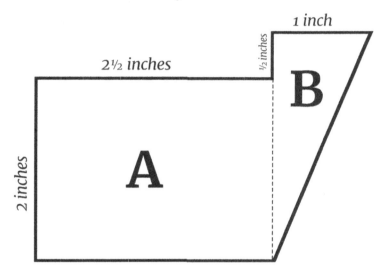

1 inch

½ inches

2½ inches

B

2 inches

A

   Measure and find the dimensions of the rectangle A. How many square yards does it contain?

3. What are the dimensions of B? How many square yards does B contain?

4. How many square yards in A and B together?

5. What part of A is B?

6. Draw a square on the scale of ¼ inch to the yard containing the same number of square yards as A+B. How many inches is each side of the square?

7. Make a rectangle like A. Into how many right triangles can you divide it? What is the area of each?

8. How many times do you breathe in 15 seconds if you breathe once every 3 seconds?

9. How many times does a person who breathes once every 4 seconds breathe in 1 minute? How long can you hold your breath?

10. Since 6 quarts of air pass into and out of the lungs every minute, how many gallons of air are used by one person in 6 minutes?

11. What part of the air is oxygen? (See Lesson 15). How many quarts of oxygen are breathed by one person every 5 minutes?

12. As only about ¼ of the oxygen breathed is utilized, how many pints of oxygen are used by one person in 5 minutes?

13. About what part of air that has been breathed is carbonic acid gas? (See Lesson 15).

14. A passenger train leaves Chicago at 4 o'clock pm, and arrives in Galesburg at 7:55 pm. How many hours and minutes does it take to make the run?

15. Where are the two hands of the clock at 4 pm and at 7:55 pm?

16. If the train leaves Galesburg 5 minutes after its arrival, how many times does the minute hand of the clock go around from the time the train leaves Chicago till it leaves Galesburg?

17. Over what part of the circle does the hour hand move in the same time?

18. A camel can travel 50 miles in 10 hours. What is its rate of travel per hour?

19. How many degrees above freezing is temperate heat? (See Lesson 16).

20. 20° above zero is how many degrees below temperate heat?

21. Name the months that have 30 days each. How many months have 31 days? How many days has February?

22. A man left home August 1 and returned October 15. How many days was he away from home?

23. How many pounds of solid material in 1¾ bushels of potatoes? (See Lesson 17).

24. If 9 pounds of every bushel of wheat is waste, how many pounds of solid food material does each bushel contain?

25. How many pounds of solid food material in 2 bushels of peas or beans? (See Lesson 17).

26. The bones of the trunk are as follows: 24 ribs, 24 vertebrae, 4 pelvic bones, 1 breastbone, and 1 bone at the root of the tongue. How many bones in the entire trunk?

27. How many bones are there in both upper limbs?

28. How many bones in both lower limbs?

29. How many more bones are there in both lower limbs than in the trunk?

30. How many more bones are there in the head than in one lower limb? (See Lesson 10).

31. About ¹⁄₁₀ of the weight of the body is bones. How much do the bones weigh in a 70-pound boy?

32. Write the multiplication table of the 5's and 6's, then repeat from memory.

> **NOTE**
> To test the accuracy of the pupils, have them take paper and pencil, then assign a problem which is to be silently read, and the answer noted on the paper. Give ten or fifteen problems on various pages, allowing a reasonable time for the silent solution of each. What percent of the answers are correct?

DRILL

1. How many quarter inches are there in 2½ inches?
2. 10 x 8 = _____
3. Ten times ½ of 4 = _____
4. Twenty is what part of 80?
5. Six times 6 divided by 4 = _____
6. One fifth of 30 quarts equals _____ quarts.
7. Six times 2 pints : 4 equals _____ pints.
8. Seven hours and 55 minutes less 4 hours equals _____ hours.
9. ⅓ of 12 = _____
10. 56° less 20° equals _____ degrees.
11. What part of 75 cents is a quarter of a dollar?
12. What do 7 boxes of strawberries cost at 7 cents a quart?
13. If 2 bananas cost 5 cents, how much do 6 bananas cost?
14. How much do 8 pounds of sugar cost at 6¼ cents per pound?
15. When beans are worth 40 cents per peck, what is the price per quart?
16. When apples sell for 60 cents a bushel, what is a peck worth?
17. How many pounds of rice may be bought for 45 cents at the rate of 9 cents per pound?

Instructions:

1. Add column *a* in 15 seconds, beginning first at the bottom and then at the top.
2. Do the same with each of the other columns.

a	b	c	d
3	1	1	4
4	4	2	2
5	3	3	3
6	6	4	4
6	6	5	3
6	6	6	4
6	5	7	5
6	1	7	8
6	4	7	8
6	6	7	8
6	7	7	8
+ 6	+ 7	+ 7	+ 8

✓ $6 \times 7 + 6 =$ _____

$7 \times 8 - 4 =$ _____

$6 \times 8 : 3 =$ _____

$7 \times 9 - 3 =$ _____

$6 \times 9 + 10 =$ _____

$8 \times 8 - 30 =$ _____

$7 \times 7 + 6 =$ _____

$8 \times 9 : 22 =$ _____

$7 \times 8 - 6 : 5 + 6 =$ _____

$\frac{1}{2}$ of $16 \times 9 + 3 : 25 =$ _____

$\frac{1}{3}$ of $15 \times 9 - 5 : 4 =$ _____

$81 : 9 \times 6 - 4 : 10 =$ _____

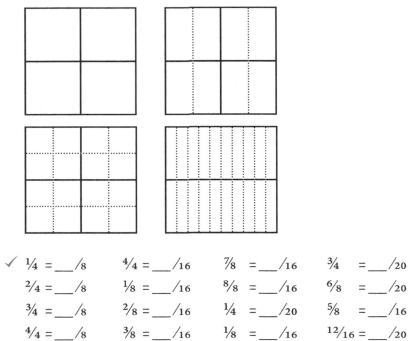

✓ ¼ = __/8	4/4 = __/16	7/8 = __/16	¾ = __/20
2/4 = __/8	⅛ = __/16	8/8 = __/16	6/8 = __/20
¾ = __/8	2/8 = __/16	¼ = __/20	5/8 = __/16
4/4 = __/8	3/8 = __/16	⅛ = __/16	12/16 = __/20
¼ = __/16	4/8 = __/16	4/16 = __/20	7/8 = __/16
2/4 = __/16	5/8 = __/16	2/4 = __/20	8/8 = __/20
¾ = __/16	6/8 = __/16	8/16 = __/20	¼ = __/40

✓ **A NUMBER EXERCISE**

Let one pupil sit facing the class and with his back toward the blackboard. If the lesson for the day is the table of 6's, as the teacher writes a number on the board, a member of the class silently multiplies this number by 6 and states the result. The pupil who is facing the class then states the number which was written by the teacher. Thus, the teacher writes the figure 5. The class says 30, and the pupil facing the class says 5. When a mistake is made by the one who cannot see the board, another takes his place, and the exercise continues.

LESSON 18 AREAS

1. Draw a rectangle 6 inches long and 2 inches wide. Mark the upper left-hand corner *a*, the lower left-hand corner *b*, the upper right-hand corner *c*, and the lower right-hand corner *d*.

   On the line *ac* measure 1½ inches from *a*, and mark *e*. On the same line, measure 1½ inches from *c*, and mark *f*.

   Connect *eb* by a line, and *fd* by a line. Mark the figure *aeb*, A. What do you call the figure A? Mark the figure *befd*, B, and the figure *fcd*, C.

   The following is the figure drawn thus far. The scale is ½ inch to the inch.

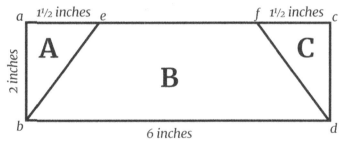

2. Cut out the rectangle, and cut off the right triangles A and C. Now, place the triangles A and C upside down upon B, so that the sides *ab* and *cd* meet, and the lines *ae* and *cf* make one straight line. What have you made?

3. How many sides of this triangle are equal? Such a triangle is called an isosceles triangle; that is, a triangle having two legs or sides equal; from Greek *isos*—equal, and *skelos*—leg.

4. How many square inches does the rectangle contain? How many square inches does the isosceles triangle contain?

5. Make a rule for computing the area of an isosceles triangle.

6. What part of the rectangle is the isosceles triangle?

7. After cutting off the isosceles triangle, what is the area of the remaining figure *befd*?

8. In this manner, compute the are of the gable of a house which is 15 feet wide, and whose peak is 6 feet above the eaves.

9. Make another figure like *befd*.

10. Can you compute the area of this figure with the following suggestion: From *b*, on the side *bd*, measure off 1½ inches, and mark *g*. Connect *e* and *g* with a line. What is the area of the figure *befd*?

11. Make another figure like *befd*. From *d*, on the side of *db*, measure off a distance equal to the side *ef*, and mark *g*. Connect *eg* with a line. Tear off the triangle, and you have left a figure whose opposite sides are equal, and therefore parallel. Such a figure is called a *parallelogram*. The suffix "gram" signifies line, from Greek *gramma*—a line.

12. Compute its areas with the following suggestions: From *d*, on the side *dg*, measure off 1½ inches, and mark *h*. Connect *f* and *h* with a line. Tear off the triangle, and place it in another position. How many square inches does the parallelogam *efdg* contain?

13. A triangular field has a base 20 yards long, and the other two side are equal. The altitude of the triangle is 9 yards. What is the area of the field?

14. From a right triangle having a base of 2 inches and an altitude of 4 inches, can you make two isosceles triangles, one with a base of 2 inches and an altitude of 2 inches, and the other with a base of 4 inches and an altitude of 1 inch?

15. A garden in the form of a rectangle contains 80 square yards, and is 10 yards long. How wide is it?

16. How long a fence would it take to enclose it?

17. If the fence posts were set 2 yards apart, how many would it take?

18. How much would they cost at 25 cents a post?

19. How many boards 12 feet long would it take to go around the fence once?

20. If the fence were 3 boards high, how many boards would be required?

21. How much would the boards cost at 10 cents each?

22. If 6 ten-penny nails were used in each board, can you tell how many ten-penny nails would be used?

23. A piece of land in the form of an isosceles triangle has a base 9 rods long. The altitude of the triangle is 18 rods. How many square rods are there in the piece of land?

DRILL

1. Six multiplied by ½ of 4 = _____
2. 15 x 3 = _____
3. Two times 4½ = _____
4. ½ of 9 multiplied by 20 = _____
5. What will 8 pounds of dates cost at 9 cents per pound?
6. How many pounds of prunes can I buy for 80 cents at 10 cents per pound?
7. When wheat is worth 80 cents per bushel, what is the price per peck?
8. How much will 9 yards of gingham cost at 9 cents a yard?
9. A boy having 3 quarters and a dime bought a straw hat for 4 dimes and a knife for a quarter. How many cents had he left?
10. A horse traveled 60 miles in 12 hours. What was the rate per hour?
11. How many degrees colder is 12° below zero than 56° above zero?
12. John lives 31 miles east of Chicago, and Mary lives 24 miles west. How far apart do they live?
13. How many dimes are there in 90 cents?
14. A half dollar, a quarter, a dime, and a nickel are _____ cents.
15. 80 : 10 = _____   Two 10's plus two 8's = _____

    36 : 2 = _____   18 x 5 = _____   36 : 4 = _____

    3 + 9 = _____   27 x 10 = _____   9 x 9 = _____

SIGHT EXERCISES

✓ Add:

4	3	5	3	6	5	7	8
6	5	6	6	7	6	6	6
3	4	4	5	8	7	3	2
7	7	8	9	9	12	13	14
6	7	8	9	11	12	13	14
+ 6	+ 7	+ 8	+ 9	+ 11	+ 12	+ 13	+ 14

3	1	2	3	4	7	6	7
4	3	4	6	7	8	4	5
5	5	6	5	8	6	4	6
6	7	8	9	11	12	13	14
6	7	8	9	11	12	13	14
6	7	8	9	11	12	13	14
+ 6	+ 7	+ 8	+ 9	+ 11	+ 12	+ 13	+ 14

✓ Use the devices given on page 53 for drill in multiplication, division, and subtraction.

✓ Write the multiplication table for the 6's, 7's, 8's, and 9's.

Ones or Units	1 Ten Ten	2 Tens Twenty	3 Tens Thirty	4 Tens Forty	5 Tens Fifty	6 Tens Sixty	7 Tens Seventy	8 Tens Eigthy	9 Tens Ninety	10 Tens One Hundred
0	10	20	30	40	50	60	70	80	90	100
1	11	21	31	41	51	61	71	81	91	
2	12	22	32	42	52	62	72	82	92	
3	13	23	33	43	53	63	73	83	93	
4	14	24	34	44	54	64	74	84	94	
5	15	25	35	45	55	65	75	85	95	
6	16	26	36	46	56	66	76	86	96	
7	17	27	37	47	57	67	77	87	97	
8	18	28	38	48	58	68	78	88	98	
9	19	29	39	49	59	69	79	89	99	

Read across the page—10, 20, 30, 40, etc.; 1, 11, 21, 31, 41, etc.

Read downward—1, 2, 3, 4, etc.; 10, 11, 12, etc.

11 equals 1 ten and 1 unit; 21 equals 2 tens and 1 unit, etc.

Draw similar diagrams and fill in, commencing with 100, 200., etc.

FRACTIONS

1. How many squares are there in A?
2. If each square represents ¼, how many fourths are there in A? ⁴⁄₄=how many whole ones?
3. How many fourths are there in B?
4. ⁵⁄₄=how many whole ones, and how many fourths?
5. How many fourths are there in C?
6. ⁶⁄₄=how many whole ones, and how many fourths?
7. How many fourths are there in D?
8. ⁷⁄₄=how many whole ones, and how many fourths?
9. One dollar equals how many quarters of a dollar?
10. How many quarters in a dollar and a quarter?
11. One dollar and a half equal _____ quarters.
12. One dollar and 75 cents equal _____ quarters.
13. Divide each of the squares that we have called *fourths* into 4 equal parts. How many of these smaller parts are there in A?
14. 1   = How many sixteenths? _____
     1¼ = How many sixteenths? _____
     1²⁄₄ = How many sixteenths? _____
     1¾ = How many sixteenths? _____

LESSON 19

TEMPERATURE PROBLEMS

1. How many seconds are there in a minute? in a minute and a half?

2. If your heart beats 1½ times each second, how many times does it beat in 1 minute? How can you tell how often your heart beats?

3. When you were 1 year old, your heart beat about 2 times each second. How many times did it beat each minute? How many more times per minute did it beat then than now?

4. An old man's heart beats about 60 times a minute, and the heart of a 2-year-old child 50 times more each minute. How many times does a 2-year-old child's heart beat in 1 minute?

5. The average rate of the heartbeat in a middle-aged person is 1⅙ times each second. What is the rate per minute?

6. Each time your heart beats, it pumps about ½ of a pint of blood. How many pints does it pump in 80 beats? How many quarts is this? How many gallons?

> What effect does exercise have upon the heart? Count the heartbeats of one of the boys. Have him run around the schoolhouse, and count again. What is the difference? How does fever affect the pulse?

7. If during a fever a person's heart beats 35 more times each minute than ordinarily, how many times does it beat, the ordinary rate being 80 times per minute?

8. The temperature of the blood, in health, is $98\frac{1}{2}°$ above zero. The temperature of a person having a fever was $104\frac{5}{10}°$. How many degrees had his temperature risen on account of the fever?

NOTE
A temperature over $107°$ is very likely to prove fatal although cases where temperature had risen to $110°$ have been known to recover.

9. The normal temperature of birds in 10 degrees higher than that of man. What is their normal temperature?

Fishes and reptiles have a lower and more variable temperature, ranging from $35°$ to $80°$. They are said to be cold-blooded.

10. The carp and toad have a temperature of $51°$. How much colder is their blood than man's?

11. How many degrees is ice-cold water raised in bringing it to the boiling point?

12. The thermometer in the schoolroom should register $68°$. How does this compare with the normal temperature of the body?

13. How far above the normal temperature of man is the boiling point of water?

14. A man seldom lives if his temperature falls below $91\frac{2}{5}°$. What are the extremes of temperature which a man may have and live?

15. Rabbits die if the temperature of the body falls to $68°$. What is the difference between the death point for the rabbit and that for a man?

16. How far below normal temperature is a bath in water of 60°?

17. The normal temperature of the exterior of the body is 98½°, but that of the interior tissues register 107°. How many tenths higher is the interior than exterior temperature?

18. If a person's temperature falls to 95³⁄₁₀°, how many tenths below normal is it?

19. On an average, a man loses from his body every 24 hours a sufficient amount of heat to raise 60 pounds of ice-cold water to the boiling point. If the loss is regular, how many pounds of water could be boiled by the heat lost in 1½ days? in 8 hours?

20. How many pints of water would thus be raised from 32° to 212° in one day? Give the answer in quarts, and then in gallons.

FRACTIONS

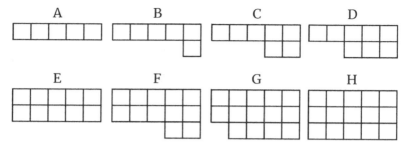

1. How many squares are there in A?
2. If each square represents ⅕, how many 5ths are there in A?
3. How many 5ths are there in B?
4. ⁶⁄₅ equal _____ whole ones, and _____ 5ths.
5. How many 5ths are there in C?
6. ⁷⁄₅ equal _____ whole ones, and _____ 5ths.
7. ¹²⁄₅ = _____   ¹⁰⁄₅ = _____

   ¹⁴⁄₅ = _____   ¹⁵⁄₅ = _____
8. If you should divide each of the squares that we have called 5ths into 2 equal parts, how many of these smaller parts would there be in A?
9. 1   = How many tenths? _____

   1⅕ = How many tenths? _____

   1⅖ = How many tenths? _____

   1⅗ = How many tenths? _____

   2⅖ = How many tenths? _____

   2⅘ = How many tenths? _____

   3   = How many tenths? _____

   4⅗ = How many tenths? _____

   5   = How many tenths? _____

   7⅖ = How many tenths? _____

REVIEW 4

1. 60 + ½ of 60 = _____   2 x 60 = _____
   60 + 50 = _____   ½ of 80 = _____
2. How many quarts are there in 40 pints?
3. How many gallons are there in 40 pints?
4. $104\frac{5}{10}°$ - 98½° equal _____ degrees.
5. 98½° + 10° equal _____ degrees.
6. 98½° - 51° equal _____ degrees.
7. 12 times 1¾ bushels equal _____ bushels.
8. 12 times 3⅓ bushels equal _____ bushels.
9. 24 : 8 = _____   ⅓ of 24 = _____
10. How many quarters are there in 1 dollar?
11. How many half dollars are there in 1½ dollars?
12. What would 8 pounds of raisins cost at 9 cents per pound?
13. If I paid for them with a dollar bill, how much change should I receive?
14. The amount of change is what part of the amount paid for the raisins?
15. Three quarters and 2 dimes equal _____ cents.
16. How many 2-cent postage stamps can be bought for 36 cents?
17. How may eleven apples be divided equally between 2 boys?
18. Henry lives 1¾ miles north of the schoolhouse, and Willie 1½ south. How far apart do they live?

✓ **Instructions:**

1. Add column *a* in 15 seconds, beginning first at the bottom and then at the top.
2. Do the same with each of the other columns.

a	b	c	d	e
1	2	1	2	1
3	4	3	4	3
5	6	6	6	5
6	7	7	8	7
6	7	6	2	9
6	7	7	4	1
6	7	6	6	3
6	7	7	8	5
6	7	7	8	7
6	7	6	8	9
6	7	6	8	9
6	7	7	8	9
6	7	7	8	9
6	7	6	8	9
+ 6	+7	+7	+8	+9

✓ Count by 10's to 100. How many tens are there in 100? How many 50's in 100?

Count by 100's to 1000. How many 100's are there in 1000? What is the difference between 100 years and 600 years? between 10 hundred years and 500 years?

◖))) ORAL EXERCISES

✓ $7 \times 8 + 10 - 6 : 4$ = _____

$6 \times 8 + 12 - 10 : 5$ = _____

$5 \times 9 - 5 : 8 + 16$ = _____

$6 \times 7 + 3 : 9 - 5$ = _____

$7 \times 9 - 3 - 40 : 5$ = _____

$8 \times 9 + 3 : 25 + 8$ = _____

$7 \times 7 + 1 - 25 : 5$ = _____

½ of $16 \times 7 + 4 - 35$ = _____

⅕ of $20 \times 8 + 3 : 7$ = _____

$9^2 - 1 : ¼ + 5 - 16$ = _____

⅓ of $24 \times 6 - 8 : 8$ = _____

$8^2 + 6 : 2 : 7 + 17$ = _____

¼ of $36 \times 6 - 4 : 2$ = _____

$2 \times 4 \times 8 - 4 : 4$ = _____

PROBLEMS IN CHRONOLOGY

> *Chronology*: from the Greek *kronos*—time, and *logia*—speak, tell, gather, read.

1. How many months are there in 1 year? in 8 years? in 10 years?

2. How many seasons are there in 8 years? How many seasons have you lived?

> What is meant by the AD, when we say this is the year 1900 AD?

3. How many hundred years ago was Jesus Christ was born?

> What is meant by the BC, when we write 600 BC?

4. Daniel was carried captive into Babylon about 600 BC. How many hundred years ago was that?

5. Abraham, the father of the Hebrew nation, was born about 20 hundred (2000) years BC, 5 hundred years later, the Hebrews were delivered from Egypt. When did the deliverance from Egyptian bondage take place?

6. Solomon's temple was built about 4 hundred years before Daniel was carried captive into Babylon. How many hundred years BC was Solomon's temple built?

7. About how many hundred years was it from the birth of Abraham to the building of Solomon's temple?

8. About how many hundred years before the building of Solomon's temple were the children of Israel delivered

from Egypt? Who was the leader of the children of Israel when they were delivered from Egypt? Exodus 13:21.

9. Ezra lived about 10 hundred years after the children of Israel were delivered from Egypt. How many hundred years BC did he live?

> Name the important events that took place about the following dates: 500 BC, 1000 BC, 1500 BC, and 2000 BC.

10. The first book of the Old Testament was written by Moses, and the last one in the New Testament by John about 100 AD. About how long a period intervened between the writing of the first and last books of the Bible?

11. About how many hundred years ago was the book of Genesis written? the book of Revelation?

12. Moses lived in Eqypt 40 years, tended flocks 40 years, and taught the children of Israel 40 years. How old was he when he died?

13. Moses was 12 years old when taken from home into the court of Pharaoh. What fraction of his entire life did he spend at home?

14. As the average length of man's life is now about 75 years, how many years longer than the average man did Moses live?

15. The flood came about 350 years before Abraham was born. How many hundred years BC was the flood?

16. At the time of the flood, the luxuriant tropical vegetation was buried beneath earth and stones. This vegetation has since turned into coal. How many hundred years ago did the trees grow that are now burned in the form of coal?

17. The earth was created more than 16 hundred years before the flood came (about 1650 years). How many hundred years BC was the world created?

18. Eijah was a prophet in Israel 100 years after King Solomon reigned. How old was he when seen with Christ on the Mount of Transfiguration?

19. Make an original problem comparing the ages of Moses and Elijah.

 DRILL

1. 1 year = _____ months
2. 2½ years = _____ months
3. 8 years = _____ months
4. 10 years = _____ months
5. Eight times 4 seasons equal _____ seasons.
6. 6 hundred years and 19 hundred years are _____ hundred years.
7. One hundred years equal one century. How many centuries ago did Abraham live?
8. 20 hundred years less 5 hundred years equal _____ hundred years.
9. 600 BC + 400 BC equal _____ hundred years BC.
10. 2000 – 1000 = _____   1500 – 1000 = _____
11. How many years from 1500 BC to 100 AD?
12. Three 40's = _____   120 – 35 = _____
13. 20 hundred + 3½ hundred = _____
14. 23½ hundred + 16½ hundred = _____
15. Add by 50's from 50 to 1000.
16. How many 50's are there in 10 hundred?

17. How many 25's are there in 5 hundred?

18. What is the area of a rectangle which is 8 feet long and 6 feet wide?

19. Draw this rectangle, allowing 2 inches to represent 1 foot. What are the dimensions in inches of the figure?

7 / VII ROMAN NOTATION

1. In the Roman notation, letters are used.

   | I = 1 | V = 5 | X = 10 |

2. The numbers from 1 to 10 are written as follows:

I	II	III	IV	V	VI	VII	VIII	IX	X
1	2	3	4	5	6	7	8	9	10

3. Placing X in succession before each of the foregoing gives the numbers from 11 to 20.

4. Give in Roman numerals:

   11 = _____   16 = _____

   12 = _____   17 = _____

   13 = _____   18 = _____

   14 = _____   19 = _____

   15 = _____   20 = _____

5. Read the following:

   XXI   = _____   XXV   = _____

   XXII   = _____   XXIV = _____

   XXVI   = _____   XXIII = _____

   XXIX   = _____   XXVII = _____

   XXVIII = _____   XXX   = _____

6. Give in Roman numerals:

31 = _____   36 = _____

32 = _____   37 = _____

33 = _____   38 = _____

34 = _____   39 = _____

35 = _____   40 = _____

L = 50   XL = 40

7. Read the following:

XLI   = _____   XLIV   = _____

XLIII   = _____   XLVI   = _____

XLV   = _____   XLIX   = _____

XLII   = _____   XLVII = _____

XLVIII = _____   L   = _____

8. Give in Roman numerals:

51 = _____   56 = _____

52 = _____   57 = _____

53 = _____   58 = _____

54 = _____   59 = _____

55 = _____   60 = _____

LX = 60   70 = _____   80 = _____

9. Read:

XXIV   = _____   LXI   = _____

XXXVI = _____   LXXIV = _____

XXXIV = _____   LXXXI = _____

XLIV   = _____   LXIX   = _____

LIV   = _____   LIX   = _____

C = 100   XC = 90

10. Give in Roman numerals:

23 = _____   63 = _____
34 = _____   78 = _____
46 = _____   82 = _____
59 = _____   94 = _____

11. Practice giving the chapter numbers in the Bible.

 SIGHT EXERCISES

✓ Add:

7	17	27	37	57	97
+ 8	+ 8	+ 8	+ 8	+ 8	+ 8

2	12	22	42	82	92
+ 7	+ 7	+ 7	+ 7	+ 7	+ 7

✓ 12   inches   make   _____   _____ .
32   quarts   make   _____   _____ .
16   ounces   make   _____   _____ .
64   pints   make   _____   bushel.
9   feet   make   _____   yards.
36   inches   make   _____   yards.
_____ inches   make   4   feet.

✓ What number multiplied by itself = 36? _____
What number multiplied by itself = 25? _____
What number multiplied by itself = 49? _____
What number multiplied by itself = 16? _____

PROBLEMS IN CUBES

1. Make a paper box after the following model. This box contains 4 cubic inches.

   Cut out the corner pieces, fold on the dotted lines, and paste the corner pieces over the joinings.

Cut out		Cut out
Cut out		Cut out

2. What is the area of the bottom of this box? What is the depth of the box?

3. If you should pack the box level full of damp sand, and then empty it on the desk, allowing the sand to retain the form of the box, how many sides of the solid figure of sand would have an area of 4 square inches? What are the dimensions of each of the other four sides?

4. Make 8 1-inch cubes after the same plan as the diagram.

5. Place two 1-inch cubes in a row. Place another row of 1-inch cubes in front of the first row. How many rows of 1-inch cubes are there?

6. Place the remaining four 1-inch cubes on top of the first four. What are the dimensions of the cube just made?

7. How many 1-inch cubes in a 2-inch cube?

8. If the 2-inch cube that you built were hollow, how many cubic inches of sand would it hold?

9. How many 1-inch cubes does it take to fill the cubic figure described in example 1?

10. Make a box that will hold:

6 cubic inches.   12 cubic inches.

8 cubic inches.   16 cubic inches.

10 cubic inches.   27 cubic inches.

11. How many cubic inches does a box contain that is 4 inches long, 2 inches wide, and 2 inches deep?

12. How wide is a box which contains 27 cubic inches, and is 3 inches long and 3 inches deep?

13. If the box, instead of being 3 inches each way, measures 3 *feet* each way, how many cubic feet does it contain?

3 feet   = 1 _____

27 cubic feet = 1 _____ _____

14. How many cubic feet in 1 cubic *yard?* in 2 cubic yards?

15. If your book is 5 inches long, 4 inches wide, and 1 inch thick, how many cubic inches does it contain?

16. How many cubic yards of earth must be removed in digging a cellar 5 yards long, 4 yards wide, and 1 yard deep?

17. How many cubic feet of earth would this be?

18. How much would it cost to have such a cellar dug at ¼ of a dollar per cubic yard?

19. Estimate the cost at 1 cent per cubic foot.

20. How many cubic feet of air in a room 10 feet long, 10 feet wide, and 8 feet high?

21. How many cubic yards in a corncrib 15 feet long, 12 feet wide, and 9 feet high?

 DRILL

1. How many cubic inches are there in a 2-inch cube?
2. How many cubic inches are there in a box 3 inches long, 2 inches wide, and 1 inch deep?
3. How many cubic inches in a box:

   4 inches by 2 inches by 1 inch?

   5 inches by 2 inches by 1 inch?

   5 inches by 2 inches by 2 inches?

   6 inches by 2 inches by 1 inch?

   3 inches by 2 inches by 2 inches?

   3 inches by 3 inches by 3 inches?

   4 inches by 2 inches by 2 inches?
4. A box containing 40 cubic inches, is 5 inches long and 4 inches wide. How deep is it?
5. How many cubic feet of air in a room 20 feet square and 10 feet high?
6. How many cubic feet in a stone wall 20 feet long, 3 feet high, and 1½ feet thick?
7. How many cubic inches in a brick that is 8 inches long, 4 inches wide, and 2¼ inches thick?
8. If a ten-year-old boy is able to inhale 125 cubic inches of air at one breath, what must be the dimensions of a box, in the form of a cube, to hold the same quantity of air?
9. 1 gallon, 1 quart, and 1 pint equal _____ pints.
10. How many quarts in 8 pecks? in 9 pecks?
11. How many pints in 7 gallons? in 9 gallons?
12. How many inches in 6 feet? in 7 feet? in 5 feet? in 8 feet? in 4 feet? in 9 feet?

13. How many inches in 1 yard? in 2½ feet?
14. How many seconds in ⅕ of a minute?
15. How many seconds in ⅗ of a minute?
16. How many seconds in ⅘ of an hour?
17. How many hours in 2½ days?
18. How many days in 9 weeks?
19. How many days in March, April, and May?
20. How many years from 1500 BC to 100 AD?
21. How many seasons have you lived?
22. How many square inches in 1 square foot?
23. How many square feet in 1 square yard?
24. How many cubic feet in 1 cubic yard?

✓ **Instructions:**

1. Add each column by sight in 15 seconds. Say 12, 24, etc., in the last column.

Five 6's + four 8's =
Seven 8's – four 9's =
Nine 9's + three 5's =

a	b	c	d	e
5	3	4	5	3
6	5	6	5	4
7	7	8	6	5
7	9	3	6	6
7	6	5	6	7
6	6	7	7	7
6	6	4	8	4
6	5	6	7	4
6	7	8	7	4
6	7	8	9	9
6	7	8	9	8
6	7	8	9	10
6	7	8	9	11
6	7	8	9	12
+ 7	+ 7	+ 8	+ 9	+12

LAND PROBLEMS

This diagram represents a tract of land containing 4 acres. It is drawn on a scale of ⅛ inch to the rod.

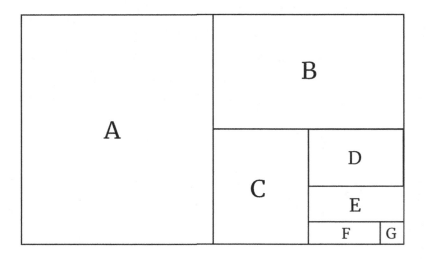

1. With a ruler, measure the figure carefully. What are the dimensions in inches of the entire 4 acres? Since this is drawn on the scale of ⅛ inch to the rod, what is the length and the breadth of the tract in rods?

2. Make this drawing on the same scale, and as each dimension is found, place the figures in the proper place.

3. What part of the 4 acres is A? What are the dimensions of A in rods?

4. A tract of land 20 rods long and 16 rods wide contains how many acres?

5. What part of A is B? What part of the 4 acres is B? What are the dimensions of B? How many square rods in B? _____ square rods make 1 acre.

6. If B were only one half as wide, how many rods long would it need to be to contain 1 acre?

7. What part of B is C? What is the area of C in square rods and acres?

   One-half acre 10 rods long is _____ wide.

8. Find the dimensions of D. What part of C is D? What part of B is D? What part of an acre is D? One fourth of 160 square rods equals _____ square rods. One fourth of 1 acre equals _____ square rods.

9. Measure E carefully. What part of D is E? What is the area of E in square rods? What part of an acre does E contain?

10. Find the dimensions and area of F. What part of E is F?

11. Find the dimensions of G.

    What part of E is G?

    What part of D is G?

    What part of C is G?

    What part of B is G?

    What part of 1 acre is 4 square rods?

1. A tract of land 20 rods long and 8 rods wide contains _____ square rods.
2. _____ square rods make 1 acre of land.
3. A 1-acre lot is 20 rods long. How wide is it?
4. A 2-acre lot is 20 rods long. How wide is it?
5. A 1-acre lot is 16 rods long. How wide is it?
6. A 1-acre plot of land is 40 rods long. How wide is it?
7. Cut a paper to show the shape of this plot.
8. What part of the length is the width of the plot mentioned in example 6?
9. 80 square rods = what part of 1 acre?
10. If ½ an acre is 8 rods in one direction, what is the other dimension?
11. If a plot of ground contains ½ an acre, and it is 20 rods long, what is its width?
12. ½ of an acre is what part of 2 acres?
    ½ of an acre is what part of 4 acres?
    ¼ of an acre is what part of ½ an acre?
    ¼ of an acre is what part of 2 acres?
    ¼ of an acre is what part of 4 acres?
    ⅛ of an acre is what part of ¼ of an acre?
    ⅛ of an acre is what part of ½ of an acre?
    ⅛ of an acre is what part of 2 acres?

✓ Add:

3	7	8	5	7	9	7	9
5	5	4	9	8	4	6	6
7	8	6	7	6	6	4	4
8	3	7	6	9	7	9	8
4	9	8	3	4	5	3	4
6	1	3	7	3	8	5	2
8	5	7	9	2	6	4	3
9	10	12	13	14	15	16	17
10	11	12	13	14	15	16	17
11	12	13	14	15	16	17	18
+ 12	+ 13	+ 14	+ 15	+ 16	+ 17	+ 18	+ 18

7	4	5	7	5	4	4	5
8	9	7	7	7	7	8	7
4	3	8	8	8	8	3	8
3	8	3	4	3	9	7	4
5	2	4	5	6	3	8	9
9	7	2	6	4	5	7	1
10	11	12	13	14	15	16	17
10	11	12	13	14	15	16	17
+ 10	+ 11	+ 12	+ 13	+ 14	+ 15	+ 16	+ 17

✓ 7 x 8 + 14 - 35 : 5 = _____   5 x 7 + 7 : 7 + 7 = _____

9 x 8 + 18 - 40 : 12 = _____   3 x 9 + 9 : 9 + 9 = _____

7 x 7 - 7 : 7 + 11 = _____   6 x 7 + 7 : 7 + 7 = _____

7 x 9 + 17 : 10 x 9 = _____   9 x 9 + 9 : 9 + 9 = _____

4 x 8 + 8 : 8 + 8 = _____   7 x 9 - 3 : 5 + 17 = _____

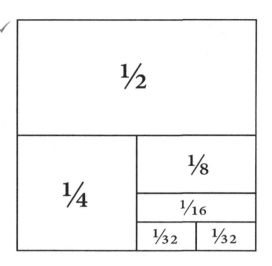

½	= how many fourths?	_____
½ + ¼	= how many fourths?	_____
½	= how many eighths?	_____
¼	= how many eighths?	_____
½+¼	= how many eighths?	_____
½+¼+⅛	= how many eighths?	_____
½-⅛	= how many eighths?	_____
⅛	= how many sixteenths?	_____
¼	= how many sixteenths?	_____
⅛+¹⁄₁₆	= how many sixteenths?	_____
¼+⅛+¹⁄₁₆	= how many sixteenths?	_____

NOTE
Use the card exercises in drilling on fractions.

REVIEW 5

1. With paper and ruler, show how to find the area of an isosceles triangle.
2. Find the area of the gable of a church 40 feet wide, the peak of the roof being 20 feet higher than the eaves.
3. Compute the area of a piece of land represented by the following figure, scale ¼ inch = 1 rod.
4. What part of an acre is represented by this figure? What part of A is B?

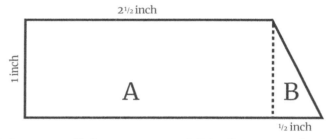

5. Make a parallelogram out of this figure, and compute its area.
6. How many square rods does your parallelogram contain? Can you make more than one kind of a parallelogram from the above figure, leaving the upper side the same length?
7. Give the difference between the number of hearbeats per minute in yourself and those in an old man.
8. How many more times per minute does a baby's heart beat than your own? How many times does your heart beat every half hour?
9. If your heart beats 80 times a minute and pumps ½ of a pint of blood at each beat, how many pints of blood does it pump in 10 minutes? How many gallons?

10. If you had a barrel holding the same quantity of water and were dipping it out with a 6-quart pail, how many dips would you have to make in order to empty the barrel?

11. How would you like to dip water in this way all day long? Does the heart get any rest at night?

12. Fever makes the heart beat about 7 more times per minute than under ordinary circumstances, for every 1 degree rise in the temperature of the body. How many more heartbeats per minute would there be in a fever patient with a temperature of 103½°?

13. How many bushels of wheat are required to sow 12 acres, if 1¾ bushels are sown to the acre? What would the seed cost at $1 per bushel?

14. How many bushels are required to sow the same field in oats, if 3⅓ bushels are sown to the acre? How much would the oats cost at 30 cents a bushel?

15. How many acres would 24 bushels of potatoes plant if 8 bushels are planted to the acre?

16. How much would the seed potatoes cost at ⅓ of a dollar a bushel?

17. How many bushels of oats are required to seed 9 acres of land?

18. How many bushels of wheat are required to sow 16 acres of land?

19. How many bushels of potatoes would you buy to plant 3 acres of land?

20. For sacking 72 bushels of wheat, 2 bushels to the sack, how many sacks are needed?

21. Compute the number of months in 5½ years.

22. How many years in 1½ centuries? How many centuries from 1000 BC to the present time?

23. Moses was 12 years of age when taken to the court of Pharaoh, and 40 when he went to the desert of Midian. How long did he live in the Egyptian court?

24. Eight furlongs make one mile. The distance from Jerusalem to Bethany was 15 furlongs. Express the distance in miles.

25. Make 3 problems in furlongs and miles using distances with which you are familiar.

26. Walking at the rate of a mile in 16 minutes, how long did it take to make the journey from Jerusalam to Bethany?

27. Bethlehem, the birthplace of David, was 6 miles from Jerusalem. Walking at the rate of 8 furlongs in 15 minutes, how long did it take to make the trip?

28. Lake Galilee averages 15 miles long by 8 miles wide. What is its area?

29. The Dead Sea averages 46 miles long by 10 miles wide. Find its area.

30. Christ visited Jerusalem at the age of 12, then returned to the carpenter's bench in Nazareth where He remained until He was 30 years of age. His ministry lasted 3½ years. How many more years did He spend in manual labor than in ministerial work?

31. How much would it cost to have a cellar dug at 25 cents a cubic yard, the dimensions of the cellar being 18 feet by 15 feet by 6 feet?

32. How many brick would it take to build a wall 10 feet long, 10 feet wide, and 1 foot thick, if there are used 20 brick to the cubic foot?

33. How much would the brick cost at ten dollars per thousand?

34. How much would the labor for laying the brick cost at 3½ dollars per thousand?

35. How many cubic feet of air are there in a room 30 feet long, 20 feet wide, and 10 feet high?

36. Allowing 600 cubic feet of air space for each person, how many persons ought to occupy the above room?

37. Make a rule for finding the area of a right triangle.

38. A flower garden is in the shape of an isosceles triangle, the base of which is 16 feet and the altitude 10 feet. What is the area of the plot?

39. If the dimensions of the triangle were rods instead of feet, how large a garden would it make? How many bushels of potatoes would be required to plant it? See example 15.

 SIGHT EXERCISES

40. Give the square of 5, 8, 10, 4, 9.

41. Give the square of 7, 6, 3, 2, 11.

42. The product of two numbers is 36, the difference 5; what are the numbers? If the product is 81 and the sum 18, what are the numbers?

LESSON 23 FRUIT-PLANTING PROBLEMS

1. Draw another plan like the one found in Lesson 22 except that E, F, and G are all in one piece, which you may mark E.

2. How many orchard trees could be set out in A if placed 2 rods apart each way, and if the rows were 1 rod from the boundary lines?

3. If the trees set out were ⅓ apple trees, ½ the remainder were peach trees, and the other ½ pear trees, how many trees of each kind would there be?

4. How much would the apple trees cost at 8 cents apiece?

5. How much would the peach trees cost at $\frac{1}{10}$ of a dollar each?

6. How much would the pear trees cost at $\frac{1}{5}$ of a dollar each?

7. If at the end of the 5th year the apple trees produced an average of 10 bushels each, what would be the total yield of apples?

8. If they were sold at 50 cents a bushel, what would the apple crop be worth?

9. If the same year the peach trees produced an average of 6 bushels each, what would be the total yield of peaches? What would be the peach crop be worth at 1 dollar a bushel?

10. If the pears averaged 5 bushels to the tree, how many bushels of pears would be raised?

11. What would the pear crop be worth at 1¼ dollars per bushel?

12. What would be the total yield in bushels of this 2-acre orchard? What would the entire crop be worth?

13. We will set out the 1-acre tract B to small fruit in rows running the long way of the piece. What will be the length of each row?

14. Place the rows 4 feet apart. Leave 2½ feet on each side. How many rows will there be?

15. If ½ of the rows are to be red raspberries, ¼ black raspberries, and the remainder blackberries, how many rows of each kind will there be?

16. If in the 3rd year the average yield of each row of red raspberries is 3 bushels, what will be the total yield of red raspberries?

17. What will they be worth at 2½ dollars a bushel?

18. What will the crop of black raspberries be worth, if they average 2½ bushels to the row and sell for 2 dollars per bushel?

19. What will the crop of blackberries be worth, if they average 4 bushels to the row, and sell for 1½ dollars per bushel?

20. What will be the total yield in bushels of this 1-acre tract of small fruit?

21. What will the total yield be worth?

22. If C is planted to potatoes, what will be the yield at the rate of 180 bushels to the acre?

23. What will the potatoes bring at ⅓ of a dollar per bushel?

We will use D for a garden, and E for the house, barn, and yards.

DRILL

1. How many trees are there in eight rows of trees, each row having 10 trees?
2. ½ of 80 = _____   ¼ of 80 = _____
3. 40 times 10 bushels equal _____ bushels.
4. ½ of 400 = _____
5. 20 times 6 bushels equal _____ bushels.
6. 20 times 5 bushels equal _____ bushels.
7. 400 bushels + 120 bushels + 100 bushels equal _____ bushels.
8. $200 + $120 + $125 equal _____ dollars.
9. 10 times 4 rows equal _____ rows.
10. ¼ of 16½ = _____
11. ½ of 40 rows equal _____ rows.
12. 20 times 3 bushels equal _____ bushels.
13. 60 times 2½ dollars equal _____ dollars.
14. ¼ of 40 x 2½ x 2 = _____
    ¼ of 40 x 4 x 1½ = _____
15. 60 bushels + 25 bushels + 40 bushels equal _____ bushels.
16. $150 + $50 + $60 = _____
    ½ of 180 = _____
    ⅓ of 90 = _____
17. How many ounces are there in 2 pounds?
18. How many ounces are there in 3½ pounds?

    10 dollars make an eagle.

19. How many eagles are there in $80?
20. 50 cents are what part of an eagle?

21. How many nickels are there in one dollar?

22. How many nickels are there in 2 dollars?

23. How many quarters in a 5-dollar bill?

✓ Add the following, up and down, and by lines to the right and left, *at sight.*

1	1	2	2	3	3	4	4	5 = ___
2	2	2	4	4	4	6	6	6 = ___
3	3	3	5	5	5	7	7	7 = ___
4	5	6	6	7	8	7	6	5 = ___
4	4	8	8	7	6	5	4	3 = ___
5	5	5	6	6	6	7	7	7 = ___
6	6	7	7	8	8	9	9	10 = ___
7	7	7	9	9	9	11	11	11 = ___
+ 7	+ 8	+ 8	+ 8	+10	+10	+10	+12	+12 = ___

✓ Use devices given on page 53 for subtraction, multiplication, and division.

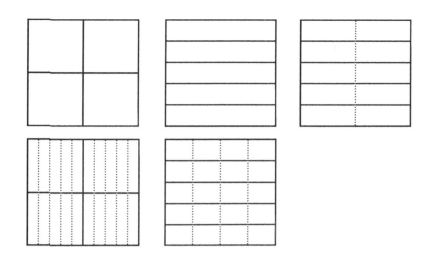

$\checkmark$ $^4/_4$ = ___ $/_5$ $^2/_5$ = ___ $/_{20}$ $^5/_{10}$ = ___ $/_{20}$

$^4/_4$ = ___ $/_{10}$ $^3/_5$ = ___ $/_{20}$ $^4/_{10}$ = ___ $/_{20}$

$^2/_4$ = ___ $/_{10}$ $^4/_5$ = ___ $/_{20}$ $^1/_4+^1/_5$ = ___ $/_{20}$

$^1/_4$ = ___ $/_{20}$ $^1/_{10}$ = ___ $/_{20}$ $^1/_4+^1/_{10}$ = ___ $/_{20}$

$^1/_5$ = ___ $/_{10}$ $^1/_5$ = ___ $/_{10}$ $^1/_5+^1/_{10}$ = ___ $/_{20}$

$^1/_5$ = ___ $/_{20}$ $^2/_5$ = ___ $/_{10}$ $^4/_5+^1/_{10}$ = ___ $/_{10}$

$^2/_4$ = ___ $/_{20}$ $^3/_5$ = ___ $/_{10}$ $^1/_4+^1/_4+^6/_{20}$ = ___ $/_{20}$

$^3/_4$ = ___ $/_{20}$ $^4/_5$ = ___ $/_{10}$ $^1/_4+^1/_5+^1/_{10}+^1/_{20}$ = ___ $/_{20}$

US MONEY

10 cents make 1 dime.
10 dimes make 1 dollar.
10 dollars make 1 eagle.

FRACTIONS & TONS

1. The composition of the human body is about as follows: ⅘ muscle and fat, ¹⁄₁₀ blood, and the remainder bones. What part is bones?

2. How much blood is there in a person weighing 120 pounds? 160 pounds?

3. How many pints of blood in the person weighing 120 pounds? How many quarts in the person weighing 160 pounds?

4. The blood is distributed to the different parts of the body as follows: ¼ to the heart, lungs, large arteries, and veins; ¼ to the liver; ⁶⁄₂₀ to the muscles and small organs; and the remainder to the brain. How much goes to the brain?

5. How much more is distributed to the liver than to the brain?

6. If there are 10 pounds of blood in your body, and ⅕ of that amount goes to your brain every 20 seconds, how many pounds of blood does your brain use in ¼ hour?

7. The average weight of the human brain is 48 ounces. What is its weight in pounds?

8. If a man's heart weighs 12 ounces, what part is that of the weight of his brain?

9. The cerebrum (Latin—the brain) fills the front and upper part of the skull, and comprises about 7/8 of the entire weight of the brain. How many ounces does it weigh in the average brain?

10. How many ounces does the cerebellum (*small brain, diminutive of the Latin* cerebrum—the brain) weigh? It is about the size of a small fist.

> The only animals whose brain outweighs man's are the elephant and the whale. The elephant's brain weighs 10 pounds and the whale's 5 pounds. Yet the human brain is heavier in proportion to its bulk.

11. $7/10$ of the weight of the body is water. How many pounds of water are there in a person weighing 100 pounds?

12. After the water has performed its work in the system, $1/5$ of it is exhaled by the lungs, $1/2$ is discharged $1/5$ by the kidneys and intestines, and the remainder by the skin. What part is discharged by the skin?

13. If we use, on an average, 100 ounces of water each day, how much of this is exhaled by the lungs?

14. How many ounces are discharged in the form of perspiration in 2 days? How many pounds in 1 day?

> 20 hundred pounds = 1 ton

15. How many pounds in $1/4$ of a ton?

16. If a farmer hauls $1\frac{1}{2}$ tons of timothy hay at each wagonload, how many tons can he haul in 5 loads?

17. If one ton of timothy hay when stowed in the haymow occupies a space of 500 cubic feet, how many tons of hay would there be in a haymow 30 feet long, 20 feet wide, and 10 feet deep?

18. If one ton of wild hay occupies a space of 400 cubic feet, how many tons could be put into the above haymow?

19. A cubic yard of sand (27 cubic feet) weighs about 3000 pounds. If this amount is hauled at one wagonload, how many wagonloads will it take to haul 3 tons of sand? How many cubic yards would this be?

20. A common horse weighs about 1000 pounds. How many horses would it take to weigh 4 tons?

21. How many 100-pound boys would it take to weigh a ton? What part of a ton would 4 boys weigh?

 **FRACTIONS**

1/4	1/4	1/4	1/4
¹/₂₀	¹/₂₀	¹/₂₀	¹/₂₀

1/5	1/5	1/5	1/5	1/5

ADDITION OF FRACTIONS

1. ¼ = how many twentieths? _____
2. ²/₄ = how many twentieths? _____
3. ¹/₅ = how many twentieths? _____
4. ¼+¹/₅ = how many twentieths? _____
5. ²/₄+¹/₅ = how many twentieths? _____
6. ¼+²/₅ = how many twentieths? _____
7. ¾+¹/₅ = how many twentieths? _____

SUBTRACTION OF FRACTIONS

8. $^4/_{20} - ^3/_{20}$ = _____
9. $^2/_4 - ^1/_{20}$ = _____
10. $^1/_4 - ^1/_5$ = _____
11. $^1/_2 - ^1/_4$ = _____
12. $^{15}/_{20} - ^1/_2$ = _____
13. $^1/_2 - ^1/_5$ = _____

MULTIPLICATION OF FRACTIONS

14. $^1/_4 \times ^1/_4$ = _____
15. $^2/_4 \times ^2/_4$ = _____
16. $^1/_4 \times ^1/_2$ = _____
17. $^1/_5 \times ^3/_4$ = _____
18. $^1/_4 \times ^2/_1$ = _____
19. $^2/_5 \times ^3/_1$ = _____
20. $^4/_5 \times ^3/_4$ = _____
21. $^2/_4 \times ^1/_2$ = _____

DIVISION OF FRACTIONS

22. Divide $^1/_2$ of an apple equally between 2 boys. What part of the whole apple has each boy?
23. Divide $^2/_4$ of an apple equally between 2 boys. What part of the whole apple has each boy?
24. $^2/_4 : 2$ = _____
25. $^1/_4 : 4$ = _____
26. $^3/_4 : 2$ = _____
27. $^2/_4 : 4$ = _____
28. $^1/_5 : 2$ = _____
29. $^1/_2 : 2$ = _____

30. $\frac{4}{5} - \frac{1}{10}$ = how many tenths? _____

31. $\frac{10}{10} - \frac{1}{10}$ = how many tenths? _____

32. $\frac{1}{10}$ of 120 = _____

   $\frac{1}{4} + \frac{1}{4} + \frac{6}{20} = \frac{9}{20} +$ _____

33. $\frac{16}{20}$ = how many fifths? _____

   $\frac{1}{5}$ of 10 = _____

34. 48 ounces = how many pounds? _____

35. $\frac{1}{8}$ of 48 = _____

   $\frac{7}{8}$ of 48 = _____

   48 – 42 = _____

36. $2\frac{1}{3}$ times 30 feet = how many feet? _____

37. 8 times 20 cents = how many cents? _____

38. 4 times 160 cents = how many cents? _____

39. $\frac{1}{4}$ of 20 hundred pounds = how many pounds?
   _____

40. $1\frac{1}{2}$ times 5 loads = how many loads? _____

41. 30 x 20 x 10 = _____

42. 60 hundred divided by 5 hundred = _____

43. How many 3000 pounds are there in 3 tons? _____

44. What part of a ton is 500 pounds? _____

45. How much will 500 pounds of coal cost at 8 dollars per ton? $_____

46. What must I pay for 1500 pounds of coal at 8 dollars per ton? $_____

47. 100 pounds are what part of a ton? _____

48. 50 pounds are what part of a ton? _____

49. 200 pounds are what part of a ton? _____

✓ Instructions:

1. Add each column in 15 seconds.

a	b	c	d
1	1	1	3
3	2	1	4
5	3	2	5
7	4	2	6
7	5	3	7
7	6		8
7	7	4	9
7	8	5	9
7	8	6	9
7	8	7	10
7	8	9	10
7	8	9	10
7	8	9	11
7	8	9	11
7	8	10	12
7	8	11	12
+ 7	+	+12	+

✓ COMMON FRACTIONS

1. How many half pecks in 36 quarts?
2. How many half gallons in 36 quarts?
3. How many half dollars in 1 eagle?
4. How many half pounds in 48 ounces?
5. How many 1/4 dollar in $5?
6. How many feet in 7 yards?
7. How many inches in 8 feet?
8. How many square inches in 1 square foot?
9. How many square feet in 8 square yards?
10. How many cubic feet in 2 cubic yards?

11. How many seconds in 2½ minutes?

12. How many minutes in 2⅓ hours?

13. How many hours in 3 days?

14. How many days in 2⅐ weeks?

15. How many weeks and days in January?

16. How many seasons in 10 years?

17. How many months in 7 years?

18. How many centuries in 1500 years?

19. How many centuries since Christ was born?

20. How many tons in 8000 pounds?

21. What part of 3 pecks is 1 quart?

22. What part of 2 gallons is 1 quart?

23. What part of 1 eagle is a half dollar?

24. What part of 4 dollars is a quarter?

25. What part of 1 square yard are 3 square feet?

26. What part of 1 cubic yard are 3 cubic feet?

27. What part of 1 minute are 15 seconds?

28. What part of 1 hour are 20 minutes?

29. What part of 1 day are 6 hours?

30. What part of 1 week are 5 days?

31. What part of 1 school month is 1 week?

32. What part of 3 years is 1 month?

33. What part of 16 centuries are 200 years?

34. What part of 1 ton are 400 pounds?

LESSON 25 PROBLEMS IN CHRONOLOGY

For the following problems, the teacher and pupils will need Bibles.

1. Twenty units make a score. How many are threescore?
2. How old was Abraham when he left Haran? (See Genesis 12:4). How many scores of years old?
3. How many years after Abraham's departure from Haran was Isaac born? (See Genesis 21:5).
4. Sarah was 10 years younger than Abraham and died when she was a hundred seven and twenty years old. How long did she live after the birth of Isaac? (See Genesis 17:17 and 23:1).
5. How old was Isaac when Abraham died? (See Genesis 25:7).
6. Isaac was threescore years old when Jacob was born. How old was Jacob when his grandfather, Abraham, died? (See Genesis 25:26).
7. How long did Jacob live? (See Genesis 35:28).
8. Shem, the son of Noah, was 98 years old at the time of the flood and lived 502 years after the flood. How old was he when he died?
9. Shem was 390 years old when Abraham was born. How long did he live after the birth of Abraham? (See Genesis 11:10-26).
10. How long did Shem live after Abraham's death?
11. Since Abraham was 100 years old when Isaac was born, how long did Shem live after the birth of Isaac?

12. Since Isaac was threescore years old when Jacob was born, how long did Shem live after the birth of Jacob?

> What opportunity had the people living in the time of Abraham, Isaac, and Jacob to learn about the flood?

13. About ⅓ of ordinary beefsteak is solid food material, and the balance is waste. How much solid food material would there be in 3 pounds of beef?

14. When beefsteak costs 15 cents a pound, how much does the solid food material in it cost per pound?

15. About ⅔ of ordinary wheat bread is solid food material. How much solid food material would there be in 3 pounds of bread?

16. When bread is worth 5 cents a loaf (1 pound), how much does the solid food material in it cost per pound?

17. At these prices, how much cheaper is it to buy bread than to buy beefsteak?

18. About ⅚ of peas and beans is solid food material. How much solid food would there be in 6 pounds of peas and beans?

19. When beans are worth 5 cents a quart, how much will 6 quarts cost?

20. How many quarts of solid food material would there be in the 6 quarts of beans?

21. How much, then, would the solid food material in beans cost per quart?

22. If one quart of beans equals 2 pounds, what is the solid food material worth per pound?

23. At these prices, how much cheaper are beans or peas than beefsteak?

> What other advantages have bread, peas, and beans over the beefsteak?

DRILL

1. 20 x 3 = _____

   100 – 75 = _____

   100 – 10 = _____

2. 127 years – 90 years = _____ years

3. 100 + threescore + 15 = _____

   175 – 100 = _____

4. 75 – threescore = _____

5. 100 years + fourscore years = _____ years

6. 98 years + 502 years = _____ years

7. 600 years – 390 years = _____ years

8. 210 years – 175 years = _____ years

9. 210 years – 100 years = _____ years

10. 110 years – 60 years = _____ years

11. ⅓ of 3 = _____   3 x 15 : 1 = _____

12. ⅔ of 3 = _____   3 x 5 : 2 = _____

13. How many 7½'s are there in 45?

14. ⅚ of 6 = _____   4 x 6 = _____

15. If 5 quarts of the solid food in beans costs 30 cents, what will 1 pound cost? If 2 pounds cost 6 cents, what will 1 pound cost?

16. 45 cents are how many times 3 cents?

17. 200 cents + 250 cents = _____ cents

18. 300 cents + 125 cents = _____ cents

19. 400 cents – 250 cents = _____ cents

20. How many half dollars are there in 250 cents?

21. How many half dollars are there in 400 cents?

22. How many quarter dollars are there in 175 cents?

✓ Add the following, up and down, and by lines to the right and left.

1	2	2	2	3	3	4	4	5 = ___
2	2	3	3	4	5	5	6	6 = ___
3	3	3	4	4	5	6	7	7 = ___
4	4	5	5	6	6	7	7	8 = ___
5	6	6	7	7	8	8	9	9 = ___
6	6	7	7	8	8	9	9	10 = ___
7	7	8	8	8	9	9	10	10 = ___
8	8	9	9	10	10	11	11	12 = ___
+ 9	+ 9	+10	+10	+10	+ 11	+ 11	+12	+12 = ___

🗣)) ORAL EXERCISES

✓ $5 \times 8 : 2 + 5 : 5 = $ ___ $9 \times 8 + 18 : 2 : 6 = $ ___

$\frac{1}{2}$ of $20 \times 8 : 4 \times 3 = $ ___ $6^2 \times 2 : 9 \times 7 - 6 = $ ___

$7 \times 8 - 6 : 2 : 5 = $ ___ $\frac{2}{3}$ of $12 \times 7 + 4 : 4 = $ ___

$6 \times 7 - 12 : 6 \times 9 = $ ___ $\frac{3}{4}$ of $8 \times 9 - 14 : 8 = $ ___

$7 \times 9 - 13 \times 2 - 40 = $ ___ $\frac{3}{5}$ of $15 \times 9 - 1 : 10 = $ ___

$8 \times 8 - 10 : 2 : 3 = $ ___ $\frac{1}{6}$ of $18 \times 9 - 2 : 5 = $ ___

✓ 210 + 120 = ___   410 + 140 = ___

500 : 200 = ___   300 + 125 = ___

300 + 175 = ___   600 + 400 = ___

500 - 250 = ___   800 - 300 = ___

200 + 176 = ___   700 + 250 = ___

300 - 150 = ___   325 + 225 = ___

✓ Give the multiplication table of 9's.

REVIEW 6

The following diagram represents a tract of land, drawn on the scale of ¼ inch to the rod. Draw a similar one on paper.

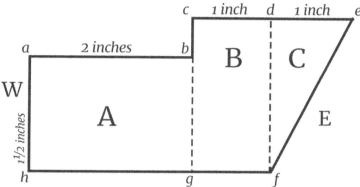

1. How many rods does the line *ab* represent?
   How many rods does the line *ah* represent?
   How many rods does the line *cd* represent?
   How many rods does the line *df* represent?
   How many rods does the line *de* represent?
2. How many square rods in A?
   How many square rods in B?
   How many square rods in C?
   How many square rods in A+B+C?
3. What part of the line *cd* is *cb*?
   What part of the line *ab* is *cb*?
   What part of the line *ah* is *cb*?
   What part of the line *hf* is *cb*?
4. How many right triangles, like C, could you draw inside of B?
5. What part of B is C? B + C = _____ square rods.

6. How many right triangles, like C, could you draw inside of A? Cut the triangle C so that it will exactly fit into the remaining portion. A = how many C's? What portion of A is C?

7. How many rods of fence will it take to enclose the entire tract, estimating the side *ef* to be 9 rods long? (It is not quite 9 rods).

8. How many more rods of fence would it take to enclose A, B, and C separately?

9. In your diagram A, make dots to show how 12 apple trees may be set out so that they shall be equally distant from each other, and the trees next to the boundary shall be equally distant from the boundary. How far apart will the trees be? How far will it be from the boundary to the first row of trees?

10. If after these trees should begin to bear well, each should produce 12½ bushels of apples. What would be the total yield for 1 year?

11. What would the crop be worth @ ⅔ of a dollar per bushel? @ 50 cents a bushel? @ 33⅓ cents per bushel?

12. If the same field A were seeded down to clover and it yielded 1500 pounds of clover, what would the clover be worth at $12 per ton?

13. How many rows of small fruit can be set out in B if 4 rows are set out to the rod, and the first row begins at C and runs north and south?

14. If each row bears 12 pecks of berries, what will the total yield be? What will they be worth at 2 dollars a bushel?

15. The tract B contains ⅕ of an acre. How many bushels of potatoes will C produce at the rate of 400 bushels to the acre?

16. What are the potatoes worth at 50 cents a bushel?

17. How much would it cost to get a team of horses shod at the rate of 20 cents a shoe? How much would it cost to get a work team shod at the rate of 2½ cents a nail? How much for a team of trotting horses at the same rate per nail?

18. If they are shod once every 3 months, what would the shoeing bill amount to for one year?

19. If a farmer hauls 1600 pounds of hay at a load, how many tons can he haul at 5 loads?

20. If a man hauls ⅔ of a cubic yard of sand at a load, how many such loads will it take to haul 2 cubic yards? About how much will each load weigh? (See Lesson 22).

21. Draw a diagram to represent the walls of a woodshed which is 10 feet square and 8 feet high. Estimate the number of square feet in the four walls.

22. In building the walls of this shed, what length boards would you buy?

23. If the boards averaged 8 inches wide, how many would it take for one side? How many for all four sides?

24. Allowing 2 feet space on the north and east inside the shed for a walk, how many cords of 4-foot wood will the shed contain filling it to the eaves?

25. If the shed has an ordinary roof with the peak 4 feet higher than the eaves, how many square feet are there in the two gables?

26. A passenger train leaves Burlington, IA at 9:05 pm and arrives in Lincoln, NE at 6:07 the next morning. How long does it take to make this run of 335 miles?

27. A steamship left New York harbor August 1 at 6 am and arrived in London, August 11 at 1 pm. How long was it on the way?

DRILL

1. How many ¼ of an inch in ½ of an inch?
2. How many ¼ of an inch in 1½ inches?
3. How many ¼ of an inch in 3 inches?
4. How many ¼ of an inch in 4½ inches?
5. On a scale of ⅛ inch to the rod, how many rods would be represented by 2 inches? by 3½ inches?
6. Draw a diagram to represent a piece of land 20 rods long and 15 rods wide on a scale of ⅛ inch to the rod.
7. What is the length and width of the diagram in inches?
8. Draw a diagram of the schoolroom showing the exact position of the teacher's desk, the stove, and other objects on a scale of ⅛ inches to the foot.
9. 12 x 12½ bushels = _____ bushels
10. ⅓ of 150 = _____
    ⅔ of 150 = _____
11. 1500 pounds is what part of a ton?
12. 3 x 17 x 2 = _____
    ½ of ⅕ of 400 = _____
13. Give the multiplication table of 8's and 9's.
14. 16 hundred times 5 : 20 hundred = _____
15. 10 x 8 x 4 = _____
16. Give the multiplication table of 10's and 11's.
17. 12 x 12 : 8 = _____

1. Add the following lines at sight in four directions.

3	3	4	4	5	5	6	6	7 = ___
4	5	5	6	6	7	7	8	8 = ___
5	6	6	7	7	8	8	9	9 = ___
6	6	7	7	8	8	9	9	10 = ___
7	8	8	8	9	9	9	10	10 = ___
8	8	9	9	9	10	10	10	11 = ___
9	9	9	10	10	10	11	12	13 = ___
10	10	11	12	12	13	13	14	14 = ___
+ 11	+ 11	+12	+12	+13	+13	+14	+14	+15 = ___

2. Use the device given on page 53 for drill in multiplication and subtraction.

> 33⅓ cents = ⅓ of a dollar
> 20 cents = ⅕ of a dollar
> 12½ cents = ⅛ of a dollar

3. At 33⅓ cents, give the cost of:

24 books	33 hats
12 dozen lemons	15 pounds prunes
30 bushels potatoes	12 knives
21 yards cloth	36 chairs
27 bushels apples	60 baskets

4. At 20 cents, give the cost of:

10 dozen oranges	40 pans
25 pecks pears	35 dosen buttons
15 knives	50 combs
20 bushels tomatoes	45 dosen lemons
30 spelling books	60 books

5. At 12½ cents, give the cost of:

8 pounds apricots	48 yards gingham
24 yards percale	64 cans peaches
32 dozen eggs	72 plates
16 writing tablets	56 dozen bananas
40 pounds prunes	80 gallons kerosene

✓ $\frac{1}{5} + \frac{3}{10}$ = __ /10

$\frac{1}{4} + \frac{4}{12}$ = __ /10

$\frac{1}{3} + \frac{2}{9}$ = __ /9

$\frac{3}{10} - \frac{1}{5}$ = __ /10

$\frac{9}{12} - \frac{1}{4}$ = __ /12

$\frac{1}{5} + \frac{2}{10}$ = __ /15

$\frac{1}{3} + \frac{1}{12}$ = __ /12

$\frac{1}{4} + \frac{1}{5} + \frac{1}{20}$ = __ /20

$\frac{5}{20} - \frac{1}{5}$ = __ /20

$\frac{7}{20} - \frac{1}{4}$ = __ /20

LESSON 26

PROBLEMS IN DECIMALS

Decimal: from the Latin *decem*—ten.

1. Divide a line into 10 equal parts. What is each part called?

2. Divide each of these parts into 10 equal parts. How many of these parts would there be in the whole line? What would each part be called?

3. If 1 hundredth of the line be divided into 10 equal parts, what would each part be called? How many of these parts in the whole line?

4. What is $\frac{1}{10}$ of $\frac{1}{10}$?
   What is $\frac{3}{10}$ of $\frac{1}{10}$?

5. What is $\frac{1}{10}$ of $\frac{1}{100}$?
   What is $\frac{7}{10}$ of $\frac{1}{100}$?

6. What is $\frac{1}{10}$ of $\frac{1}{1000}$?
   What is $\frac{1}{100}$ of $\frac{1}{100}$?

7. What part of $\frac{1}{10}$ is $\frac{1}{100}$?

8. What part of 1 tenth is 25 hundredths?

The following fractions are all decimal fractions:

$\frac{1}{10}$, $\frac{3}{10}$, $\frac{1}{100}$, $\frac{5}{100}$, $\frac{12}{100}$, $\frac{1}{1000}$, $\frac{13}{1000}$, $\frac{216}{1000}$

How do these decimal fractions differ from the following common fractions?

$\frac{2}{3}$, $\frac{3}{5}$, $\frac{4}{8}$, $\frac{5}{12}$, $\frac{6}{140}$, $\frac{245}{3176}$, etc., etc.

When we write $256.37, the 2 stands for _____ dollars, the 5 means _____ dollars, the 6 means _____ dollars. The 3 stands for 3 times what part of a dollar? The 7 stands for 7 times what part of a dollar?

When the dollar mark is omitted, 256.37 is read two hundred fifty six and thirty seven hundredths.

9. The distance around a square is 20.32 feet. What is the length of each side?

10. How long will it take a man to walk 21.27 miles at the rate of 3 miles per hour?

11. If 5.6 bushels of potatoes be equally divided among 7 families, what part of a bushel would each family receive?

12. A boy earned .35 of a dollar for 7 hours' work. How much did he get an hour?

13. If he earned .45 of a dollar the next day, how much money did he then have?

14. James had $7.35 and spent $1.15. How much money did he then have?

15. James had $2.15 and spent $1.05. How much did he have left?

16. A man bought a bed for $6.50, a chair for $1.25, and a mirror for $1.00. How much money did he spend?

17. A man worked 5 days at $1.20 a day. How much did he earn?

18. What will 3 barrels of flour cost at $4.25 per barrel?

19. If it costs $1.25 to lay 1 thousand shingles, what will it cost to lay 5 thousand?

20. If a man earns $1.75 each day, how much will he earn in 2 days? in 6 days?

DRILL

1. What is $\frac{7}{10}$ of 100 pounds?
2. $\frac{1}{5} + \frac{1}{2} =$ _____

   $\frac{1}{10} + \frac{7}{10} =$ _____
3. What is $\frac{1}{5}$ of 100? _____ $\frac{3}{10}$ of 100? _____
4. $\frac{3}{8}$ of 8? _____

   $8 + 4\frac{1}{2} =$ _____

   3 times $\frac{3}{4} =$ _____
5. At the rate of 30 miles per hour, how far will a train go in 8 hours?
6. At 7 cents a yard, what will 20 yards of calico cost? 50 yards? 100 yards?
7. If a boy saves $8 a month, how much can he save in a year?
8. At $1\frac{1}{2}$ cents apiece, what will 5 dozen eggs cost?
9. At the rate of 3 oranges for 10 cents, what will 1 dozen cost?
10. How many pounds of coal can be bought for 1 dollar at the rate of 5 dollars per ton?
11. When hickory wood is worth $2\frac{1}{2}$ dollars a cord, how much must I pay for 7 cords?
12. If 3 tons of coal at 8 dollars per ton will last 6 months, what is the cost of coal per month?
13. If 8 pounds of apricots cost 1 dollar, what is the price per pound?
14. If 3 bushels of apples cost 1 dollar, what will 1 bushel cost?
15. What will 3 pounds of apricots cost at 12.5 cents per pound?

✓ Add the following, up and down, and by lines to the right and left, *at sight.*

4	4	5	5	6	6	7	7	8 = ___
5	6	6	7	7	8	8	8	9 = ___
6	6	7	7	8	8	9	9	9 = ___
7	7	8	8	8	9	9	10	10 = ___
8	8	8	9	9	10	10	11	11 = ___
9	9	9	10	10	10	11	11	12 = ___
10	10	11	11	11	12	12	12	13 = ___
11	11	11	12	12	13	13	14	14 = ___
+12	+12	+12	+13	+13	+14	+14	+15	+15 = ___

✓ EXERCISE IN MULTIPLICATION

1. What is the square of 4? of 5? of 6?
2. What is the square of 7? of 8? of 9?
3. What is the square of 10? of 11? of 12?
4. What will 12 pounds of sugar cost at 6¼ cents per pound?
5. Find the cost of:

   8 pounds of rice at 7½ cents   _____

   9 yards of calico at 6⅓ cents   _____

   30 quarts of berries at 5½ cents   _____

   12 gallons of milk at 20 cents   _____

   4 tons of coal at 6¼ dollars   _____

   12 tons of hay at 8⅓ dollars   _____

   6 thousand feet of lumber at 15 dollars a thousand

   _____

   10 acres of land at 25 dollars   _____

6. Find the cost of:

26 books at 50 cents   _____

24 bushels apples at 33⅓ cents   _____

48 brooms at 25 cents   _____

60 baskets at 20 cents   _____

72 dozen oranges at 12½ cents   _____

✓ 6 x 12 = _____	5 x 12 = _____
4 x 11 = _____	10 x 12 = _____
12 x 8 = _____	7 x 11 = _____
12 x 12 = _____	9 x 8 = _____
11 x 12 = _____	9 x 9 = _____
7 x 12 = _____	12 x 9 = _____
7 x 9 = _____	12 x 3 = _____
10 x 11 = _____	2 x 12 = _____
4 x 12 = _____	6 x 9 = _____

PROBLEMS IN BOARD MEASURE

Any board, one inch or less in thickness, containing a surface area of 144 square inches or one square foot, is called a *board foot.*

A board:

1 inch wide and 12 feet (144 inches) long	=	1 board foot
2 inches wide and 12 feet (144 inches) long	=	2 board foot
3 inches wide and 12 feet (144 inches) long	=	3 board foot
6 inches wide and 12 feet (144 inches) long	=	6 board foot
12 inches wide and 12 feet (144 inches) long	=	12 board foot

1. How many board feet in a board 10 inches wide and 12 feet long?

2. How many board feet in 5 boards 10 inches wide and 12 feet long?

3. A box is 6 feet long, 4 feet wide, and 3 feet high. How many square feet are there in the two sides which are 6 feet by 4 feet? How many boards 12 inches wide and 12 feet long would it take for these two sides? Into how many pieces would you saw each board?

4. How many square feet are there in the two sides which are 6 feet by 3 feet?

5. How many boards 12 inches by 12 feet would it take for these two sides? Into how many pieces would you saw each board?

6. How many square feet would there be in the two sides which are 4 feet by 3 feet? How many boards would it take for these two sides, and how would you saw them?

7. How many square feet are there in the six sides of the box, and how many boards 12 inches wide and 12 feet long, will it take to make the box?

8. A man left home June 1 and returned home July 14. How many days was he away from home?

9. A passenger train leaves Chicago at 5:15 pm and arrives in Omaha at 8:15 am the next day. How many hours does it take to make the run?

10. The distance from Chicago to Omaha is 500 miles. What will the railroad fare be at 3 cents a mile?

11. A passenger train leaves Chicago Monday at 4 pm and arrives in Denver at 6:30 pm Tuesday. How many hours does it take to make the run?

12. The distance from Chicago to Denver is 1000 miles plus 25 miles over the CB & QRR. What will the railroad fare be at 3 cents per mile?

13. This same train leaves Denver at 7 pm Tuesday and arrives in San Francisco at 11:45 am Friday. How many hours does it take to make the run from Denver to San Fransisco?

These trains do not actually run to San Francisco, but to Oakland pier on the eastern shore of San Francisco Bay about five miles from the city.

14. How many days and how many hours does it take this train to run from Chicago to San Francisco?

15. A passenger train on the Rock Island route leaves Chicago Monday at 10 pm and arrives in San Francisco

Friday at 4:15 pm. How long does it take this train to make the trip?

16. The distance from Chicago to San Francisco is 26 hundred miles plus 15 miles. What is the railroad fare at 3 dollars per hundred miles?

17. From Chicago to San Diego, CA, it is 3000 miles by railroad. What would the trip cost at 3 cents a mile? What would it cost at 2½ cents a mile?

18. A passenger train leaves Chicago at 2 pm and arrives in New York at 6 pm the following day. How long does it take to make the run?

19. The distance from Chicago to New York by railroad is 1000 miles. What is the fare at 2 cents a mile?

20. How far is it by railroad from New York to San Diego?

21. How many days will it take to go from New York to San Francisco?

DRILL

1. Estimate the number of board feet in the following:
   1 board 4 inches wide and 12 feet long
   3 boards 4 inches wide and 12 feet long
   5 boards 6 inches wide and 12 feet long
   10 boards 10 inches wide and 12 feet long
   10 boards 12 inches wide and 12 feet long
   7 boards 8 inches wide and 12 feet long
   9 boards 8 inches wide and 12 feet long

2. If a board 1 inch wide and 12 feet long contains 1 board foot, how many board feet would there be in a board 1 inch wide and 6 feet long?

3. How many board feet in:
   1 board 4 inches wide and 6 feet long
   2 board 6 inches wide and 6 feet long
   4 board 6 inches wide and 6 feet long
   10 board 12 inches wide and 6 feet long
   100 board 10 inches wide and 12 feet long

4. What is the difference in time between 5:15 pm and 8:15 am the following day?

5. 500 times 3 cents = _____ cents? How many dollars?

6. How many hours from 4 pm Monday to 6:30 pm Tuesday?

7. 1000 x 3 cents = _____ cents

8. How many hours from 7 pm Tuesday to 11:45 am Friday?

✓ Add the following, up and down, and by lines to the right and left.

2	2	2	3	3	3	4	5	6 = ___
3	3	3	4	4	5	6	7	7 = ___
4	4	4	6	6	6	7	8	8 = ___
5	6	7	7	8	8	8	9	9 = ___
6	6	6	7	7	7	8	8	8 = ___
7	7	7	8	8	8	9	9	10 = ___
8	8	8	9	10	10	10	11	11 = ___
9	9	10	10	11	11	12	12	12 = ___
+10	+10	+ 11	+ 11	+10	+12	+12	+13	+13 = ___

✓ Use the following device for drill in multiplication and subtraction as in Review 1.

$$-13$$

1	2	3	4	5	6	7	8	9	10	11	12

$$7$$

> 50 cents = ½ of a dollar
> 25 cents = ¼ of a dollar
> 10 cents = ¹⁄₁₀ of a dollar

✓ At 50 cents per pound, yard, dozen, etc., what will be paid for:

1. 24 books? _____
2. 12 dozen oranges? _____
3. 30 pounds bromose? _____
4. 36 yards cloth? _____
5. 50 bushels apples? _____
6. 32 bushels tomatoes? _____
7. 18 bushels potatoes _____
8. 48 panels of glass? _____
9. 38 hats? _____
10. 45 yards cashmere? _____

✓ At 25 cents, give the cost of:
1. 8 knives   _____
2. 12 dozen lemons   _____
3. 28 yards cloth   _____
4. 25 spelling books   _____
5. 32 pairs hinges   _____
6. 40 meals   _____
7. 45 pair scissors   _____
8. 48 combs   _____
9. 60 baskets   _____
10. 61 cans of paint   _____

✓ At 10 cents, give the cost of:
1. 20 pounds prunes   _____
2. 30 pounds rice   _____
3. 40 tablets   _____
4. 70 yards cloth   _____
5. 100 dozen bananas   _____
6. 75 dozen eggs   _____
7. 45 quarts raspberries _____
8. 120 pecks potatoes   _____
9. 80 pounds nut butter _____
10. 95 pounds dates   _____

✓ What is the tithe on 35 dollars? on 50 cents? on 78 dollars? on 10 cents?

✓ Pupils should bring in ten original problems.

BOARD MEASURE

1. Common boards may be brought at the lumber yard in lengths of 10, 12, 14, or 16 feet. A board 14 feet long contains ⅙ more board feet than a 12-foot board, the other dimensions being equal. Show by a diagram that this is true.

2. How many board feet in:

   1 board 12 inches wide and 12 feet long?

   1 board 12 inches wide and 14 feet long?

   10 boards 6 inches wide and 14 feet long?

3. What length and width boards would you buy if you wished to make the top of a table 4 feet wide and 7 feet long in order that there may be as little waste as possible?

4. How much would the lumber cost at 2 cents per board foot?

5. A man wishes to make a fence 3 boards high around a garden which is 84 feet long and 70 feet wide. If the posts are set 7 feet apart, how many will it take?

6. How many boards 6 inches wide and 14 feet long will it take for one side which is 84 feet long? How many for both sides?

7. How many boards will it take for one end which is 70 feet long? How many for both ends?

8. How many will it take altogether?

9. How many board feet does each board contain?

10. How many board feet will there be in all the boards which he buys?

11. When the lumber sells for $20 per thousand feet, what is the price per hundred feet?

12. What is the price per foot?

13. What will the man's fence boards cost him at this price?

14. What will the posts cost at 10 cents each?

15. The distance from New York to Chicago in a straight line is 7 hundred miles; from Chicago to Denver, 9 hundred miles; from Denver to San Francisco, 9½ hundred miles. How far is it from Chicago to San Francisco by way of Denver?

16. How far is it from New York to Denver by way of Chicago? How far is it from New York to San Francisco by way of Chicago and Denver? What is the difference between this distance and the railroad distance given in the Lesson 27?

17. A steamboat left Boston harbor Monday, September 3 at 6 am, and arrived in Liverpool harbor September 12 at 12 pm. On what day of the week did it arrive?

18. How many days and hours did it take to make the trip?

From a map of the world, find how far it is from Boston to Liverpool.

19. A man traveled by wagon from Omaha to Denver, a distance of 500 miles, in 28 days. Allowing 3 days of rest, what was the rate of travel per day?

20. A passenger train runs 380 miles in 10 hours. What is its average speed per hour?

21. A hawk is able to fly 125 miles in an hour. How far can it travel in 10 hours? How far could it travel in 2 days, flying 10 hours each day? in 20 days?

> The answer to the last example is the distance around the world.

22. A camel has four stomachs, in one of which it can store a large quantity of water, so that it is able to go without drink for 12 days. It is able to travel 50 miles a day. How far can it travel without taking a drink?

> It is able to do this with a 600-pound load on its back.

 DRILL

1. 1⅙ times 12 board feet = _____ board feet?
2. 4 x 7 divided by 14 = _____
3. 84 : 7 + 1 x 2 = _____   70 : 7 –1 x 2 = _____
4. 84 : 14 x 3 x 2 = _____   70 : 14 x 3 x2 = _____
5. 36 + 30 + 4 = _____
    70 x 7 = _____   44 x 10 = _____
6. 7 hundred + 9 hundred + 9 1/2 hundred = _____ hundred?
7. 900 + 700 = _____   45 x 8 = _____
8. 900 + 950 = _____   3650 – 2550 = _____
9. 28 days – 3 days = _____ days
10. 500 : 25 = _____   380 : 10 = _____
11. 125 x 10 = _____
    1250 x 2 x 10 = _____   12 x 50 = _____

12. How many board feet in:

10 boards 6 inches wide and 12 feet long?

10 boards 6 inches wide and 14 feet long?

10 boards 12 inches wide and 14 feet long?

10 boards 14 inches wide and 12 feet long?

10 boards 14 inches wide and 14 feet long?

10 boards 11 inches wide and 12 feet long?

20 boards 4 inches wide and 6 feet long?

13. What will 200 feet of lumber cost at $20 per thousand?

14. What will 200 feet cost at $30 per thousand?

15. What will 200 feet cost at $20 per thousand?

✓ Add the following, up and down, and by lines to the right and left, *at sight.*

3	4	4	5	5	6	6	7	7 = ___
5	5	6	6	7	7	8	8	8 = ___
5	6	6	7	7	8	8	9	9 = ___
6	7	7	8	8	9	9	10	10 = ___
7	7	8	8	9	9	10	10	11 = ___
8	8	9	9	10	10	11	11	12 = ___
8	9	9	9	10	10	11	12	13 = ___
10	10	11	11	12	12	13	13	14 = ___
+ 11	+ 11	+12	+12	+13	+13	+14	+14	+15 = ___

✓ Use other devices already given for drill in multiplication, division, and subtraction.

✓ Find the cost of:
1. 72 books at 50¢ _____
2. 32 yards cloth at 12½¢ _____
3. 24 bushels corn at 33⅓¢ _____
4. 12 dozens plates at 50¢ _____
5. 24 dozens lemons at 25¢ _____
6. 36 bushels apples at 33⅓¢ _____
7. 48 gallons gasoline at 12½¢ _____
8. 10 dozens bananas at 20¢ _____
9. 200 pounds prunes at 10¢ _____
10. 45 chairs at 50¢ _____
11. 100 dozen oranges at 25¢ _____
12. 72 gallon kerosene at 12½¢ _____
13. 33 yards cloth at 33⅓¢ _____
14. 80 boxes granose at 12½¢ _____
15. 150 pounds apricots at 10¢ _____
16. 75 books at 20¢ _____
17. 27 bushels potatoes at 33⅓¢ _____
18. 62 cans tomatoes at 10¢ _____
19. 60 bushels tomatoes at 33⅓¢ _____
20. 60 pounds raisins at 12½¢ _____

THE ERECTION OF A SMALL SCHOOL BUILDING

FIGURE A

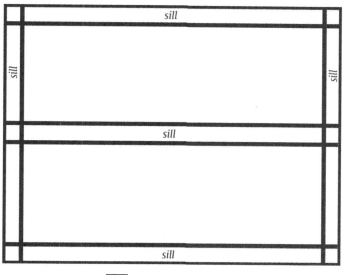

☐ End of sill

1. Draw a rectangle 4 inches by 5 inches. From each corner, measure on the line ⅛ of an inch each way, and mark with a dot. Connect these dots by four lines parallel to the sides and ends of the rectangle, making an inner rectangle. Mark the center of each end of the outer rectangle. On the line, ¹⁄₁₆ of an inch each way from these center dots, make dots. Connect these dots by lines parallel to the sides of the rectangle.

2. On the scale of ¼ inch to the foot, what are the dimensions of our house whose sills are represented by the parallel lines ⅛ of an inch apart?

3. On the foregoing scale, how many inches wide are our sills?

4. $\frac{1}{12}$ of an inch will represent the thickness of the sills. Draw a rectangle representing the end of one of them on the same scale.

5. How wide and how thick are the sills of the house?

6. How many board feet in the three sills which are parallel?

> **NOTE**
> Make a bill of the lumber beginning as follows: 3 sills 6 x 4 inches, 20 feet long x 120 board feet.

7. How many feet in the two end sills? How many board feet in the 5 sills just mentioned?

8. On the sides of the inside rectangle, make dots $\frac{3}{8}$ of an inch apart, commencing at the left. Connect the opposite dots by lines parallel to the end sills. Draw lines parallel to each of these lines $\frac{1}{24}$ of an inch from them.

FIGURE B

9. These double lines are floor joists. They are each 2x10 inches and 16 feet long. How many of these joists did you draw?

10. How many board feet in each joist? How many in the 14 joists?

11. Draw a rectangle 2½ inches by 5 inches. This represents one side of the house. How high is it?

FIGURE C

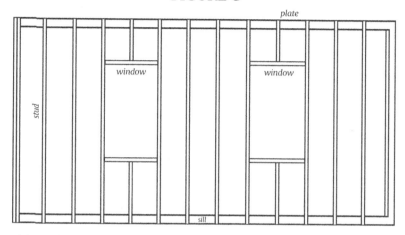

12. ¹⁄₁₂ of an inch below the lower side of the rectangle, draw a line parallel to it. If ¼ inch=1 foot, what will ¹⁄₁₂ inch equal?

13. What does the rectangle ¹⁄₁₂ inch wide and 5 inches long represent? Here the thickness of the sill can be seen. This could not be seen on the other plan. How long and how thick is this sill?

14. On the 5-inch lines (Figure C), make dots ⅜ of an inch apart. Connect opposite dots (except the fifth dots from each end), and ¹⁄₂₄ of an inch from each of these lines draw a parallel line. At each corner, draw two parallel lines.

15. These double lines are called studs. Each stud is 2x4 inches and 10 feet long. At the corner, there are two nailed together making it 4x4 inches. How many studs are there?

16. On the fourth stud from each end, place a dot ⅝ of an inch above the sill. One and one-half inches above each of these dots, place a dot. From these dots, draw double lines parallel to the sill across the two wide spaces between the studs. What have you made?

17. What are the dimensions in feet of the spaces left for windows? How many feet from the top of the sill to the bottom of the window space?

18. ¹⁄₂₄ of an inch above the top side of the rectangle (Figure C), draw a line parallel to it, thus representing what is called the plate.

19. How far apart are the studs, except the first two at one end?

20. How many board feet in the 14 long studs?

21. How many board feet in the short pieces above and below the window spaces?

22. How many feet in the plate which is 2x4 inches, 20 feet long?

23. What is the total amount of board feet in the framework of this side of the house? Enter this on your bill.

24. The opposite side of the house has the same dimensions as this one.

25. Follow these instructions.

   a. Draw a rectangle 4 inches long and 2⅝ inches high.

   b. One twelfth of an inch above the lower side, draw a line parallel to it.

   c. Mark the center of the upper side.

   d. From this mark, measure 1½ inches perpendicular to the top side, and make a dot.

e. For the moment, imagine that your rectangle is a map, and make a dot ½₄ of an inch northeast of the upper right-hand corner of the rectangle, and another dot ½₄ of an inch northwest of the upper left-hand corner.

f. From the dot made under *d*, draw two lines to the dots under *e*, projecting beyond the dots ⅜ of an inch.

g. Below and ½₄ of an inch from the two lines last drawn, draw parallels. Do not cross the corner of the rectangle.

h. One twenty-fourth of an inch from the projections spoken under *f*, draw lines parallel to them.

i. One twelfth of an inch inside each end of the rectangle, draw lines parallel to the ends.

j. Erase the upper side of the rectangle.

FIGURE D

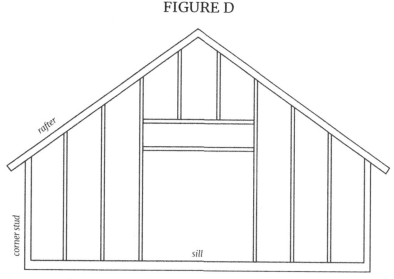

26. Answer these questions.

a. What have we now represented?

b. What is the distance from the top of the sill to the peak of the roof?

c. How much higher is the peak of the roof than the top of the side plate?

d. What is the distance from the top of the sill to the top of the side plate?

e. How wide are the rafters which form the roof (excepting the projections)?

f. How wide are the projections?

g. How wide are the corner studs?

27. Follow these instructions.

a. Divide the still into 3 equal spaces and mark. From the two marks made, draw double lines to the rafters parallel to the corner studs.

b. The center space is for a double door.

c. On each side of the door, draw 2 double lines to represent studs, dividing the space equally.

d. On each of the door studs, 2¼ inches above the sill, put a mark.

e. Connect the two dots by a double line.

f. Three eighths of an inch below this line, draw another double line parallel to it.

g. The space for between these two double lines is for what?

h. Above the transom, draw two double lines representing studs.

28. Follow these instructions, and solve the problems.

a. In estimating the board feet in the framework of this end, why not figure on the sill and two corners?

b. Leave out the two rafters also, and figure on them with the other rafters.

c. Measure the 6 long studs, and remember that studding comes in lengths of 10, 12, 14, and 16 feet.

d. How many of the studs must be cut from 14-foot material?

e. How many from 12-foot material?

f. From what shall we cut the two short studs and pieces over the door and transom to avoid waste?

g. Our bill then for this end will be:

4 studs 2x4 inches, 14 feet long=_____ board feet

4 studs 2x4 inches, 12 feet long=_____ board feet

29. In estimating the material for the other end of the house, leave out the door space, and the bill will be as follows:

2 studs 2x4 inches, 16 feet long=_____ board feet

4 studs 2x4 inches, 14 feet long=_____ board feet

2 studs 2x4 inches, 12 feet long=_____ board feet

30. What is the total number of board feet in the framework of both ends? How many more in the back than in the front of the building?

31. The ceiling joists will be the same length as the floor joists, and 2 inches by 4 inches.

32. On the top of the plate, make dots ½ inch apart (Figure C). These dots show where the ends of the ceiling joists rest upon the plate. How many will there be? How far apart will they be? The two end ones, however, must be placed in from each corner 2 inches to make room for the end roof rafters.

33. What fraction of an inch will represent 2 inches on the scale of ¼ inch=1 foot?

34. Referring to Figure D, two sets of roof rafters are seen, the upper ends uniting at the top of the plates. (Notice how they are sawed at each end). The rafters are made from timber 2x4 inches. Measure them. How long is the timber used in making them?

35. The rafters are placed the same distance apart as the ceiling joists. How many are there on each side?

36. Find the number of board feet in the rafters.

30 PERCENTAGE

Percent means by the hundredths. 5 percent means 5 one hundredths; 10 percent means 10 one hundredths, etc. The symbol of percent is %.

> 50 cents = $^{50}/_{100}$ or ½
> 25 cents = $^{25}/_{100}$ or ¼
> 20 cents = $^{20}/_{100}$ or ⅕

1. If in canning 32 quarts of strawberries 50 percent is lost in waste and shrinkage, how many 1-quart jars can be filled?

2. Allowing 50 percent for waste and shrinkage in canning blackberries, how many gallons of berries will it take to put up 16 2-quart jars?

3. If the berries cost 12 cents a gallon, what would each quart of berries be worth when canned, allowing 3 cents a quart for labor, fuel, etc.?

4. Allowing 20 percent loss for waste and shrinkage in canning plums, how many 1-quart jars can be put up from 20 quarts of plum? How many from 50 quarts? from 100 quarts?

5. A lady canned 40 quarts of strawberries and not having sealed them tightly, 25 percent of them spoiled. How many quarts had she remaining?

6. From a certain amount of strawberries, a lady put up 10 quarts. If 50 percent was lost in waste and shrinkage, what was the amount of fruit used?

7. If 20% is lost in canning plums, what percent of the original bulk is canned?

8. If in canning 40 quarts of fruit, 10 quarts are lost in waste, what is the percent of waste?

9. If 8 quarts of fruit are canned from 4 gallons, what percent is lost by shrinkage?

10. If in canning 32 quarts of fruit, 8 quarts are lost in shrinkage, what percent is the shrinkage?

11. If 40 quarts of fruits are canned from 50 quarts, what percent is the shrinkage?

12. 20% of the air is oxygen. How many cubic feet of oxygen are there in a room 10 feet high and 10 feet square?

13. In breathing 100 cubic feet of air, how much oxygen is used by the system if 20% is oxygen, and only 25% of the oxygen taken into the lungs is used?

$33\frac{1}{3}$ percents	$= \frac{1}{3}$
$12\frac{1}{2}$ percents	$= \frac{1}{8}$
10 percents	$= \frac{1}{10}$

14. $33\frac{1}{3}$ percent of the bones in an adult person is animal matter, and the remainder is mineral matter. How much would a 12-pound bone weigh after being burned?

15. If 10 percent of the weight of the body is bones, how many pounds of mineral matter would there be in a person who weighed 150 pounds?

16. If 10 percent of the weight of the human body is blood, and 20% of the blood passes to the brain, how many pounds of blood go to the brain of a man who weighs 150 pounds?

17. Moses was 120 years old when he died. $33\frac{1}{3}$% of his life was spent in the desert of Midian, and 50% of the remainder was spent with the children of Israel in the

wilderness. Into what three periods may his life be divided?

18. 10% of this same man's life was spent with his mother before he entered the Egyptian court. How old was he when he went into the court of Pharaoh?

19. Nebuchadnezzar's golden image (Daniel 3) was threescore cubits high and 6 cubits broad. What percent of the height was the breadth?

20. The king's steward proved Daniel by giving him pulse to eat for 10 days. What percent of 1 month of 30 days was that? What percent of 2 months is 10 days?

NOTE

Have pupils write four things which can be seen in problem 19. Also, have them make four problems for their classmates. To illustrate, two things can be seen are:
1. The difference between the length and the breadth is 54 cubits.
2. The height is 180 feet, and the breadth 18 feet.

PROBLEMS

1. What is the difference between the height and the breadth expressed in feet?
2. 18 feet are what percent of 180 feet?
3. The breadth is what fraction of the difference between the height and breadth?
4. Put other statements similar to the one in problem 19 on the board, and the pupils will be obliged to think for themselves in following your directions.

 **DRILL**

1. What does percent mean?
2. 50 percent = _____ hundredths
3. 25 percent = _____ hundredths
4. 20 percent = _____ hundredths
5. 33⅓ percent = _____ hundredths
6. 12½ percent = _____ hundredths
7. 10 percent = _____ hundredths
8. In finding 33⅓%, what easy fraction do you use?
9. In finding 12½%, what easy fraction do you use?
10. What is 50 percent of 32 quarts?
11. 32 quarts are 50 percent of _____ quarts.
12. 96 cents : 32 = _____ cents
13. What is 20% of 20 quarts? of 50 quarts? of 100 quarts?
14. What is 20% of 8? of 16? of 80?
15. 10 quarts are 50% of _____ quarts.
16. 100% – 20% = _____
17. 10 quarts are what part of 40 quarts?
18. ¼ = what percent? _____
19. 8 quarts are what part of 16 quarts? ½ = _____%
20. 8 quarts are what part of 32 quarts? ¼ = _____%
21. 10 quarts are what part of 50 quarts? ⅕ = _____%
22. 20% of 1000 = _____

    20% of 100   = _____

    25% of 20 = _____
23. ⅓ of 12 = _____

    12 – (⅓ of 12) = _____

    10% of 150 = _____

24. 33⅓% of 15 = _____
25. 15 − (33⅓% of 15) = _____
26. What is 20% of 15 pounds?
27. What is 12½% of 48 bushels?

✓ Add the following, up and down, and by lines to the right and left, *at sight*.

4	4	4	5	5	6	6	7	7 = ___
5	5	5	6	6	7	7	8	8 = ___
6	6	7	7	8	8	9	9	10 = ___
7	8	8	8	9	9	10	10	10 = ___
8	8	9	9	9	10	10	11	11 = ___
9	10	10	10	11	11	11	12	12 = ___
10	10	10	11	11	12	12	13	13 = ___
11	11	12	12	12	13	13	13	14 = ___
12	12	13	13	13	14	14	15	15 = ___
13	13	14	14	15	15	16	16	16 = ___
+14	+14	+15	+15	+16	+16	+16	+17	+17 = ___

✓ 1. 4½ + 3¼ = _____   6. 12 + 4½ + 3¼ = _____
2. 5¾ + 2¼ = _____   7. 15 + 5¾ + 2¼ = _____
3. 7½ − 3¼ = _____   8. 17 + 6⅓ − 3⅓ = _____
4. 12²⁄₄ − 3½ = _____   9. 18 + 4⅓ + 2⅔ = _____
5. 8¾ + 1½ = _____   10. 10 + 3¾ + 1½ = _____

11. What is 50% of 6 x 6? _____ 19. 225 −125 = _____

12. What is 25% of 8 x 10? _____ 20. 246 − 216 = _____

13. What is 33⅓% of 4 x 9? _____ 21. 175 − 140 = _____

14. What is 20% of 5 x 12? _____ 22. 156 − 114 = _____

15. What is 12½% of 4 x 8? _____ 23. 215 − 125 = _____

16. What is 10% of 5 x 8? _____ 24. 280 − 235 = _____

17. What is 33⅓% of 5 x 12? _____ 25. 300 − 150 = _____

18. What is 20% of 5 x 8? _____ 26. 425 − 300 = _____

16⅔ cents = ⅙ of a dollar
8⅓ cents = 1/12 of a dollar

✓ Find the cost of:
1. 12 dozen eggs at 16⅔ cents. _____
2. 24 pounds of dates at 8⅓ cents. _____
3. 15 yards of cloth at 33⅓ cents. _____
4. 40 pans at 25 cents. _____
5. 40 dozen lemons at 12½ cents. _____
6. 30 dozen bananas at 20 cents. _____
7. 24 books at 16⅔ cents. _____
8. 30 chairs at 33⅓ cents. _____
9. 36 knives at 25 cents. _____
10. 48 tablets at 12½ cents. _____

LESSON 31

PROBLEMS FROM THE FARM

A man has $100 invested in a farm. The following is a diagram of the farm drawn on the scale of ¹⁄₁₆ inch to the rod. Draw a similar plan on paper.

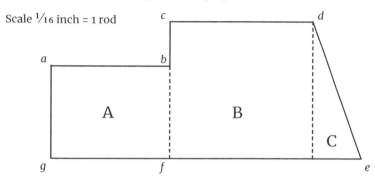

Scale ¹⁄₁₆ inch = 1 rod

1. How many rods does the line *ab* represent?
2. How many rods does the line *ah* represent?
3. How many rods does the line *cd* represent?
4. How many rods does the line *df* represent?
5. How many rods does the line *ef* represent?
6. How many square rods are there in A?
7. How many acres are there in A?
8. How many square rods are there in B?
9. How many acres are there in B?
10. How many square rods are there in C?
11. How many acres are there in C?
12. How many square rods are there in A+B+C?
13. How many acres are there in A+B+C?
14. How much did the plot cost per acre?

15. If 50% of B+C is planted in corn, how many bushels could be raised at the rate of 60 bushels to the acre?

16. What would the corn crop be worth at 33⅓ cents per bushel?

17. If the remainder of B+C is sown to wheat and clover, how many bushels of wheat could be raised at the rate of 30 bushels to the acre? (The clover does not yield until the second year.)

18. What would the wheat be worth at ⅘ of a dollar per bushel? What would be the worth of the corn and wheat crop raised from B+C in 1 year?

19. Divide the tract A into 4 equal parts, so that each part is 10 rods long and 8 rods wide. What part of an acre does each of these four tracts contain? Mark the upper left-hand tract *i*, the upper right-hand tract *j*, the lower left-hand tract *k*, and the lower right-hand tract *l*. The tract *l* will be used for a house, barn, etc. 25% of 320 square rods = _____ acres.

20. How many apple trees can be set out in the tract *j*, if they are placed 2 rods apart each way, and the trees next to the boundary are placed 1 rod from the boundary?

21. If, in 6 years from the time of setting out, each tree bears 10 bushels of apples, what is the yield from all the trees?

22. What would the crop be worth at 50 cents a bushel?

23. At this rate, how many bushels of apples could be raised on a tract of land the size of B+C?

24. What would the apple crop be worth?

25. If B+C is put in corn or wheat each year, and the worth of the crop raised averages as estimated in the

preceding problems, what would the crops for 5 years be worth?

26. How many bushels of apples could be raised in 5 years on a tract of land the size of B+C if each tree continued to bear 10 bushels of apples?

27. What would the apple crops for these 5 years be worth at 50 cents per bushel?

28. At this rate, how much more money could be made in 10 years in raising apples than in raising corn, wheat, etc., on a 3-acre tract of land?

29. How many rows of blackberries and raspberries can be set out in the tract *i* if the rows run north and south, and are set out 4 rows to the rod?

30. If each row bears, on an average, 2 bushels of berries, how many bushels can be raised in all?

31. What would the berry crop be worth at 2 dollars a bushel?

32. At this rate, what would the berry crop be worth on a piece of land the size of B+C?

33. ½ of the tract *k* is planted to potatoes, and they yield 240 bushels to the acre. What would the crop be worth at 30 cents a bushel?

34. If the other ½ of *k* is put in garden, and it yields a crop worth $50, what would 1 acre of garden yield at the same rate?

35. What % of B is C?

36. What % of A is C?

37. 50% of A is what % of B+C?

38. 25% of B is what % of A+C?

39. What % of the line *he* is the line *ab*?

1. How many 16ths of an inch in 1¼ inches?
2. How many 16ths of an inch in 3 inches?
3. 20 x 16 = _____   320 : 160 = _____
4. 20 x 20 = _____   20 x ½ of 8 = _____
5. 400 : 160 = _____   400 + 80 = _____
6. 480 : 160 = _____   100 : 5 = _____
7. ⅓ of 90 = _____   ⅘ of 45 = _____
8. ½ of 20 x 10 = _____   36 + 30 = _____
9. 5 x 66 = _____   200 x 6 = _____
10. 5 x 1200 = _____   6000+1200 x ½ = _____
11. 10 x 66 = _____   10 x 4 x 2 x 2 = _____
12. 6 x 160 = _____   8 x 25 = _____

13. A man having a 20-dollar bill, spent $12 for a suit of clothes, $1¼ for a hat, and $2¾ for a pair of shoes. What % of his money had he left?

14. What is the cost of 25 bushels of potatoes at ⅕ of a dollar per bushel?

15. What is the cost of 25 bushels of wheat at ⅘ of a dollar per bushel?

16. A grocer bought 30 gallons of syrup for $24 and sold it for 25 cents a quart. How much did he gain? What % of the cost was the gain?

17. A farmer sold 1 bushel, 3 pecks, and 4 quarts of clover seed at $8 a bushel. How much did he get for it?

✓ Add the following, up and down, and by lines to the right and left, *at sight.*

5	5	6	6	7	7	8	8	9 = ___
6	7	7	8	8	9	9	10	10 = ___
7	7	8	8	9	9	10	10	11 = ___
8	8	8	9	9	10	10	11	11 = ___
8	9	9	10	10	11	11	12	12 = ___
9	9	10	10	11	11	12	12	13 = ___
10	11	11	12	12	13	13	14	14 = ___
11	11	12	12	13	13	14	14	15 = ___
+12	+13	+13	+13	+14	+14	+15	+16	+17 = ___

🗣))) ORAL EXERCISES

✓ 1. 7 x 8 + 4 : 3 + 5 : 5   = ___
2. ⅓ of 24 x 9 – 2 : 2 : 7   = ___
3. ⅔ of 12 x 8 – 4 : 12 x 6   = ___
4. 7 x 7 + 1 : 10 x 7 : 10   = ___
5. 7 x 9 – 23 : 8 x 9 + 20   = ___
6. ¾ of 12 x 6 x 6 : 5 x 7   = ___
7. 9 x 9 – 21 : 3 + 7 : 8   = ___
8. ⅘ of 10 x 7 – 6 : 25 x 9   = ___
9. 11 x 11 – 21 : 10 x 8 : 40   = ___
10. 10 x 12 + 5 : 25 x 7 + 4   = ___

✓ 11. If ⅔ of a pound of sugar cost 4¢, what will 1 pound cost?

12. If ¾ of a pound of prunes cost 6¢, what will 1 pound cost?

13. If ⅘ of a pound of rice cost 8¢, what will 1 pound cost?

14. If ⅜ of a yard of cloth cost 3¢, what will 1 yard cost?

15. If ⅝ of an acre of land cost $5, what will 1 acre cost?

16. If ⅜ of an acre of land cost $6, what will 1 acre cost?

17. If ⅝ of an acre of land cost $10, what will 1 acre cost?

18. If ⅞ of an acre of land cost $14, what will 1 acre cost?

✓ Have the pupils pass to the board, or to take their slates, and write the multiplication tables of 6's, 8's, 7's, 9's, and 12's.

REVIEW 7

Boards used for sheeting a house, or for other common purposes, are not all the same size. They vary from 2 to 16 or more inches in width. In buying a load of lumber, the lumberman measures the width of each board put on the wagon.

1. In buying a load of sheeting 12 feet long, the lumberman's tapeline shows the total width of the boards to be 120 inches. How many feet are there?

2. If in buying boards 12 feet long, the total width of the boards is 50 feet, how many feet of lumber are on the wagon?

3. If the load is dry pine lumber, each board foot will weigh about 2½ pounds. How many pounds would then be on the load?

4. How many board feet will it take to sheet the 4 walls of a house represented by the following diagram, scale ¹⁄₁₆ inch to the foot?

5. Draw the same diagram on a scale of ¼ inch to the foot. Fold on the lines *a*, *b*, and *c*, and fasten the two ends d and e together, thus making a house.

6. What will the sheeting for the 4 walls and gables cost at $20 per thousand?

7. If the distance from the eaves to the peak of the roof (*f* to *g*) is 10 feet, how many board feet will it take to sheet the roof?

> **NOTE**
> Roof sheeting is not placed together, a space of 2 or 3 inches being left between each board; but the projection of the roof over the ends and eaves requires enough lumber to lift up these spaces.
> Shingles, as a rule, are laid 4 inches to the weather, and 1000 shingles are estimated to cover 100 square feet.

8. How many shingles will it take to cover this house?

> **NOTE**
> Each side of the roof is 10 by 20 feet. In addition to this area, there is a space 2 feet wide running all around the edges of the roof. This is the projection at the ends and eaves.

9. What is the cost of shingles at $2 a thousand? What is the cost at $3 a thousand?

10. A train left Cincinnati, Ohio, at 10 pm, and arrived in Toledo at 5:10 the next morning. How long did it take to make the run?

11. Another train left Cincinnati at 8:20 am and arrived in Toledo at 2:20 pm the same day. How much less time did it take this train to make the run than the other?

12. The distance from Cincinnati to Toledo by railroad is 203 miles. What is the fare at 2½ cents a mile?

    200 x 2½ = _____   3 x 2½ = _____

13. A steamer left Sydney, Australia, Wednesday, August 1 at 9 am for San Francisco, touching at Auckland, New Zealand, and Honolulu. It made the trip from Sydney to Auckland in 6 days and remained in Auckland 8 hours. The time from Auckland to Honolulu was 16 days, and a stop of 1 day was made at Honolulu. The time from Honolulu to San Francisco was 8 days. How many days did it take to make the trip?

14. Upon what day of the month was the arrival in San Francisco?

15. What was the day of the week and the hour of its arrival?

16. Another steamer left San Francisco, Monday, August 6 at 6 am for Yokohama, Japan, and made the trip in 17 days, 4 hours, and 30 minutes. Give the day of the month, the day of the week, the hour and minute, of its arrival in Yokohama.

17. About how much water is exhaled by the lungs in 1 day, estimating that we use 5 pounds of water each day?

18. How much would be discharged by the skin?

19. A merchant bought cloth for 60 cents a yard and sold it at a gain of 25%. How much did he make on each yard?

20. A merchant marks his goods so as to make a profit of 20%. How shall a pair of shoes be marked that cost $2.50?

21. A canvasser makes 33⅓% on each book. How much will he make in selling 5 books at 60 cents each?

22. If in 1 month a canvasser delivers 80 $1 books, how much money does he make if his commission is 50%?

23. If out of this he pays 10% tithe, 40% for board, room, etc., and 25% for other expenses, how much money has he left?

24. How many $1 books must I sell in order to make $8 a week if my commission is 50%?

25. How many $1 books must I see in order to make $8 a week if my commission is 33⅓%?

26. If a man makes $10 in selling 30 1-dollar books, what is his commission?

27. If a man sells $50 worth of books and sends $30 to the firm for which he is working, what percent commission does he get?

28. On a scale of ¼ inch to the rod, draw the boundary line of a farm which runs as follows: Beginning at A, the northwest corner, the line runs south 20 rods to B. From B, it runs east 16 rods to C. From C, it runs north 10 rods to D. From D, it runs northwest to A.

29. How many inches long is your line which represents twenty rods? How many inches long is the line which represents sixteen rods? How many inches long is the line which represents ten rods?

30. How many acres are there in this piece of land?

31. Draw a rectangle, on the same scale, which contains the same number of acres.

32. Require pupils to bring in five problems.

✏ DRILL

1. 120 inches = _____ feet   10 x 12 = _____
2. 12 x 50 x 2½ = _____
3. Two 20's plus two 16's = _____
   5 times ½ of 16 = _____
4. 16 times ½ of 5 = _____   20 x 20 = _____
5. 20 + 16 + 20 + 16 x 2 = _____
6. 25% of 60 = _____   33⅓% of 60 = _____
7. 20% of $2.50 = _____   50% of $27 = _____
8. 40% of $40 = _____   50% of $80 = _____
9. 10% of $80 = _____   25% of 80 = _____
10. 40% of 80 = _____   20% of 45 = _____

11. $8 is 50% of _____ dollars.

12. $10 is what percent of $30? _____

13. 10 is what percent of 40? _____

14. 12 is what percent of 36? _____

15. $2\frac{1}{2}$ is what percent of 5? _____

16. 2 is what percent of 16? _____

17. 40 is what percent of 100? _____

18. 20 is what percent of 50? _____

19. 5 is what percent of 25? _____

20. $12\frac{1}{2}$% of 32 = _____   20% of 200 = _____

21. $16\frac{2}{3}$% of 36 = _____   40% of 200 = _____

22. 25% of 10 = _____ $16\frac{2}{3}$% of 60 = _____

23. $33\frac{1}{3}$% of 90 = _____ $12\frac{1}{2}$% of 80 = _____

✓ Add the following, up and down, and by lines to the right and left.

7	8	9	10	10	11	11	12	12 = ___
8	9	10	10	11	11	12	12	13 = ___
9	10	10	11	11	12	12	13	13 = ___
10	10	11	11	12	12	13	13	14 = ___
10	11	11	12	12	13	13	14	14 = ___
11	11	12	12	13	13	14	14	15 = ___
11	12	12	13	13	14	14	15	15 = ___
12	12	13	13	14	14	15	15	16 = ___
+13	+13	+14	+14	+15	+15	+16	+16	+17 = ___

1. 50% of 16 x 6 + 2 -25   =_____
2. 33⅓% of 15 x 9 - 5 : 8   =_____
3. 25% of 40 x 8 -30 : 2   =_____
4. 16⅔% of 36 x 6 : 6 - 6   =_____
5. 12½% of 32 x 9 + 4 : 4   =_____
6. 5^2 x 2 x 30 : 4   =_____
7. 6^2 + 4 : 8 x 9   =_____
8. 7^2 + 11 : 3 : 4   =_____
9. 8^2 + 16 : 4 : 5   =_____
10. 9^2 + 19 - 50 : 10 =_____

11. ½ + ¾   =_____
12. ¾ + ⅜   =_____
13. 2½ + ⅛   =_____
14. 3¼ + ⅜   =_____

15. 7⅜ + ¼   =_____
16. 6³⁄₁₆ + ¼   =_____
17. 7⅕ + ¹⁄₁₀   =_____

18. 2½ + 3¼ + 4   =_____
19. 3⅕ + ¹⁄₁₀ + 11 ⅗   =_____
20. 4⅓ + 3⅔ + 7   =_____
21. 5⅕ + 4¹⁄₁₀ + 3¹⁄₂₀ =_____
22. 6⅓ - 4½ + 2¼   =_____
23. 7¹⁄₂₅ + 2¹⁄₅₀ + ½   =_____
24. 8⅓ + 3½ + 2¼   =_____

25. 12 x 2½   =_____
26. 30 x 3⅓   =_____
27. 40 x 2¼   =_____
28. 25 x 3⅕   =_____

29. 15 x 3⅔   =_____
30. 16 x 4¼   =_____
31. 20 x 5⅗   =_____

FRACTIONS

1. Mary bought ¾ of a yard of cloth at 40 cents a yard. What did she pay for it?

¼ yard 10 cents	¼ yard 10 cents	¼ yard 10 cents	¼ yard 10 cents

ANALYSIS
At 40 cents a yard, 1 fourth of a yard will cost ¼ of 40 cents, or 10 cents; and 3 fourths of a yard will cost 3 times 10 cents, or 30 cents.

2. When apples are worth 48 cents a bushel, what will 3 pecks cost?

3. What will 1500 pounds of coal cost at $10 a ton?

4. At 15 cents a dozen, what will 8 bananas cost?

5. What part of 35 days is 3 weeks?

6. When milk is worth 16 cents a gallon, what will 3 pints cost?

7. Mary spent ⅗ of a dollar for 3 yards of cloth. What was the price per yard?

8. John has a board ¾ of a foot long. If he saws off ⅓ of it, how long will the piece be that was sawed off?

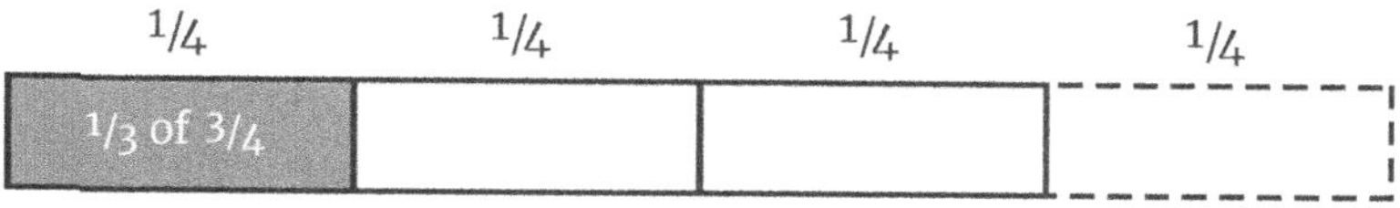

¼	¼	¼	¼
⅓ of ¾			

9. What part of the board remained? What part of a foot was it in length?

10. A man had ⅘ of an acre of land and sold ¼ of it. Show by means of a diagram what part of an acre he sold.

11. If he sold this piece at the rate of 30 dollars an acre, what did he get for it?

12. If a mother divides ⅚ of an apple equally among 5 children, what part of an apple will each child receive?

13. A canvasser sold $60 worth of books at a commission of 33⅓%. How much tithe should he pay on his commission?

14. What is ⅔ of ¾?

15. What is ⅔ of 6/7?

16. What is ¾ of ⅘?

17. What is ¾ of 8/9?

18. What is ⅘ of ⅚?

19. What is ⅘ of 10/12?

20. From a bin containing 60 bushels of wheat, 45 bushels were sold. What percent of the wheat remained?

21. What did the wheat which was sold bring at 66⅔ cents a bushel?

22. A bin which holds 40 bushels of oats is ¾ full. If ⅔ of the oats are sold, how many bushels will there still be in the bin?

23. What would the ⅔ of the ¾ of 40 bushels bring at ⅖ of a dollar a bushel?

24. A man having ⅘ of an acre of land, set out ¾ of it in small fruit, and the rest he planted to potatoes. What part of an acre was set out in small fruit?

25. What would be the yield of small fruit at the rate of 150 bushels to the acre?

26. What would the frut be worth at 2⅓ dollars a bushel?

27. What would be the yield of potatoes at the rate of 200 bushels to the acre?

28. What would the potatoes be worth at 40 cents a bushel?

29. A grocer buys eggs at 10 cents a dozen and sells them for 12 cents a dozen. What percent does he make?

30. If from 8 gallons of blackberries 28 quarts are canned, what percentage is lost in shrinkage?

31. If from 32 quarts of plums 24 quarts are canned, what percent is lost in shrinkage? What percent is canned?

32. If from 12½ gallons of grapes 20 quarts of grape juice are canned, what percent is canned?

✏ DRILL

1. What is ⅓ of ¾? _____   ¾ – ¼ = _____
2. What is ¼ of ⅘? _____   What is ¾ of ⅘? _____
3. What is ⅕ of ⅚? _____   What is ⅗ of ⅚? _____
4. What is ⅕ of 30? _____   What is ⅘ of 30? _____
5. What is 33⅓% of 60? _____
   What is 10% of 20? _____
6. What is ¾ of 40¢? _____
   What is ⅓ of 48¢? _____
7. 1500 pounds are what part of 1 ton?
8. 500 pounds are what part of 1 ton?
9. 400 pounds are what part of 1 ton?
10. 1200 pounds are what part of 1 ton?
11. 8 is what part of 12?   ⅔ of 15 = _____
12. 35 days equal how many weeks?
13. 3 pints are what part of a gallon?

14. ⅗ of 100 cents divided by 5 = _____
15. 60 – 45 = _____
    15 is what percent of $60? ___
16. ⅔ of ¾ of 40 = _____   ¾ of 8 = _____
17. ⅗ of 150 = _____   ⅘ of 200 = _____
18. 3⅓ x 60 = _____
    2 is what percent of 10? _____
19. 2 is what percent of 32? _____
    12½% of 64 = _____
20. How many books at $1 each must be sold in order to make $10 a week if the commission is 33⅓%?
21. A canvasser sold $54 worth of books and sent the firm for which he was working $36 as their share. What was his percent on the sales?

✓ Find the number of board feet in:
    1. 10 boards 10 inches wide and 12 feet long
    2. 50 boards 5 inches wide and 12 feet long
    3. 10 boards 12 inches wide and 14 feet long
    4. 20 boards 6 inches wide and 6 feet long
    5. 8 boards 12 inches wide and 12 feet long
    6. 11 boards 8 inches wide and 6 feet long
    7. 8 boards 8 inches wide and 12 feet long
    8. 8 boards 6 inches wide and 14 feet long

👓 SIGHT EXERCISES

✓ 1. 5 x 8 + 2 : 7   = _____
  2. 4 x 9 : 3 – 8   = _____
  3. 7 x 8 + 4 : 12   = _____
  4. 9 x 8 : 2 – 20   = _____

5. $6 \times 7 : 2 - 14$ = _____
6. $9 \times 9 + 9 : 2$ = _____
7. $6^2 + 4 - 8 + 16$ = _____
8. $10^2 - 25 + 5 : 4$ = _____
9. $7^2 + 14 : 7 \times 9$ = _____
10. $\frac{1}{3}$ of $36 \times 8 : 4$ = _____
11. $\frac{3}{4}$ of $16 \times 9 + 13$ = _____
12. $\frac{4}{5}$ of $20 - 4 \times 7$ = _____
13. 5 is what % of 20? = _____
14. 6 is what % of 48? = _____
15. 7 is what % of 21? = _____
16. 8 is what % of 64? = _____
17. 9 is what % of 90? = _____
18. 10 is what % of 50? = _____
19. 32 is what % of 64? = _____
20. 5 is 10% of what number? = _____
21. 6 is 12½% of what number? = _____
22. 7 is 20% of what number? = _____
23. 8 is 25% of what number? = _____
24. 9 is 33⅓% of what number? = _____
25. 10 is 40% of what number? = _____

✓ Have the pupils bring in 6 problems on addition of common fractions, 6 on subtraction of fractions, and 7 each in multiplication and division of common fractions. These problems should be original ones.

LESSON 33

SCIENCE PROBLEMS

1. There are 4 chambers in the heart, each holding about ¼ of a pint. All the blood in the body passes through each of these chambers. Since 10% of the weight of the body is blood, and 1 pint equals 1 pound, how many times would the left ventricle (one of the chambers of the heart) have to be filled and emptied in order that all the blood of a person weighing 120 pounds may pass through it?

2. The daily work of the heart is said to be equal to ⅓ that of all of the muscles of the body. Each time it beats, it exerts a force equal to lifting 5 pounds 1 foot high. How many pounds could it lift 1 foot high in a minute, counting 80 beats to the minute?

3. At this rate, how long would it take the heart to lift 1 ton 1 foot high?

4. How many tons could it lift in 1 hour?

5. If the heart pumps 24 pints of blood to the lungs each minute. How many gallons of blood go to the lungs in an hour?

6. If a person breathes once for every 4 heart beats, how many times will he breathe in 1 minute, counting 72 beats to the minute?

7. If ⅔ of a pint of air are taken in at each breath, how many pints of air are breathed in 1 minute? How many quarts is this?

8. The surface of the lungs exposed to the blood as it circulates through them, if spread out, would cover a room 28x50 feet. How many square feet is that?

9. Our lungs purify about 10 tons of blood every day. How many pounds are purified each day?

10. If you had to pump this same weight of water from a well, how many pailfuls would you have to pump if each pailful weighed 50 pounds?

11. How many hours would it take to do this, pumping 1 pailful every 3 minutes?

12. If a grocer buys 48 bushels of apples, and 12½% of them decay before he sells them, how many bushels will he have left to sell?

13. If he buys them for 33⅓ cents and sells them for 50 cents, does he make or lose, and how much?

14. A man sorted over 80 bushels of apples and found that he had 60 bushels which were sound. What percent were unsound?

15. From 1 bushel blackberries 28 quarts may be canned. What percent is lost in shrinkage?

16. The catalogue price of a book is 80 cents. If 12½% discount is allowed the purchaser, what does the book cost?

17. What is the cost of a book catalogued at 75 cents and sold at 20% discount?

18. If I buy a book for 60 cents whose catalogue price is 90 cents, what percent discount do I get?

1. What is 10% of 120? _____
    What is 5% of 120? _____
2. How many ¼ pounds are there in 12 pounds?
3. 5 x 80 = _____
    6 x 800 = _____
    2000 : 400 = _____
4. 24 pints equal _____ gallons.
5. 72 : 4 = _____
    ⅔ of 18 = _____
    ¾ of 20 = _____
6. 28 x 50 = _____   10 x 2000 pounds = _____
7. 20,000 : 50 = _____
    3 x 400 = _____
    12½% f 48 = _____
8. ⅓ of 48 = _____   ½ of (48 – 6) = _____
9. 80 – 60 is what percent of 80? _____
10. 32 – 28 is what percent of 32? _____
11. 12½% of 80 = _____
12. 20% of 75 = _____
13. A man selling goods on 33⅓% commission makes $30; what amount of goods did he sell?

    > **ANALYSIS**
    > 33⅓% = ⅓. If $30 = ⅓, ⅓ = 3 x $30 or $90.

14. A canvasser makes $40 selling books at a commission of 40%. What was the value of the books sold?
15. A grocer sells eggs at a profit of 12½% and makes 10 cents. How many cents' worth of eggs did he sell?

16. 3 is 20% of what number?

17. 4 is 33⅓% of what number?

18. 2 is 40% of what number?

19. 4 is 12½% of what number?

REVIEW OF DECIMAL & COMMON FRACTIONS

Decimal fractions are usually expressed by writing the numerators only, the denominators being indicated by the use of a dot placed before the numerator. Thus:

$\frac{1}{10}$ is written .1	$\frac{1}{1000}$ is written .001
$\frac{4}{10}$ is written .4	$\frac{24}{1000}$ is written .024
$\frac{1}{100}$ is written .01	$\frac{356}{1000}$ is written .356
$\frac{24}{100}$ is written .24	$\frac{1}{10000}$ is written .0001

The dot used in writing decimal fractions is called a *decimal point*.

Notice in the above decimals that the number of places of figures in the decimal is the same as the number of ciphers in the denominator. If, as in some of the examples, the numerator of the decimal fraction has a less number of figures than there are ciphers in the denominator, the deficiency is supplies by prefixing one or more ciphers.

Write the following fractions as decimals.

✓ 1. $\frac{8}{10}$ = _____   6. $\frac{342}{1000}$ = _____

2. $\frac{12}{16}$ = _____   7. $\frac{56}{1000}$ = _____

3. $\frac{5}{100}$ = _____   8. $\frac{2}{1000}$ = _____

4. $\frac{24}{100}$ = _____   9. $\frac{10}{1000}$ = _____

5. $\frac{6}{100}$ = _____   10. $\frac{2141}{1000}$ = _____

11. $^4/_{1000}$ = _____ 14. $^{123}/_{100}$ = _____

12. $^{55}/_{1000}$ = _____ 15. $^{45}/_{10}$ = _____

13. $^{174}/_{1000}$ = _____ 16. $^{4234}/_{1000}$ = _____

Read the following.

✓ 1. .6   6. .018   11. .0205   16. .2010

2. .13   7. .009   12. .0084   17. .104

3. .04   8. .756   13. .307   18. .0005

4. .27   9. .24   14. .20   19. .632

5. .125   10. .8   15. .1066   20. .1401

Add the following.

✓ 21.   16.4   22.   18.3   23. 325.4   24. 428.01
         3.12   5.25   4.16   35.6
       + 4.05   + 7.4   + 5.23   + 28.35

✓ 25. $^2/_{10}$ + $^3/_{10}$ = _____

.7 + .5 = _____

.8 + .4 + .5 = _____

26. .8 − .5 = _____

.12 − .03 = _____

27. What is the sum of .1, .12, and .08?

28. What is the sum of .$^3/_5$ and $^7/_{10}$?

29. What is the sum of .9, .23, and .17?

30. What is the sum of 4.21 and 3.07?

31. From 3.15, subtract .12.

32. From 4.23, subtract 2.11.

33. From 2.1, subtract 1.8.

✓ Multiply ¹⁄₁₀ by 10; by 100; by 1000.
Multiply .3 by 10; by 100; by 1000.
Multiply .9 by 100.
Multiply 3.4 by 10; by 100.

> **PRINCIPLES**
> 1. A decimal is multiplied by 10, 100, 1000, etc., by moving the decimal point as many places toward the right as there are ciphers in the multiplier.
> 2. A decimal is divided by 10, 100, 1000, etc., by moving the decimal point as many places toward the left as there are ciphers in the divisor.

✓ 1. Multiply .25 by 2.
2. Multiply .17 by 3.
3. Multiply .12 by 6.
4. Multiply $2.25 by 3.
5. Divide .24 by 6.
6. Divide .56 by 8.
7. Divide 1.44 by 12.
8. Divide .08 by 4.
9. Multiply .05 by 8.
10. Multiply .18 by 5.
11. Divide .75 by 5.
12. Divide $1.25 by .25.

✓ Express as common fractions.

1. .3 = _____	7. .352 = _____
2. .5 = _____	8. .106 = _____
3. .13 = _____	9. .25 = _____
4. .8 = _____	10. .0110 = _____
5. .16 = _____	11. .0008 = _____
6. .2 = _____	12. .0175 = _____

INTEREST

Interest is money paid for the use of money.
The principal is the sum of money which is borrowed.
The rate of interest is the percent paid for the use of the principal.

Find the interest on:

1. $100 at 6% for 1 year.

2. $100 at 6% for 6 months.

3. $200 at 6% for 1 year.

4. $200 at 6% for 2 years and 6 months.

5. $100 at 10% for 1 year and 6 months.

6. $100 at 5% for 1 year.

7. $125 at 5% for 1 year and 6 months.

8. $175 at 5% for 6 months.

9. January 1, 1900, a man gave his note for $100 payable July 1, 1900, with interest at the rate of 6% from date. How much money will he pay the person whom he borrowed the money July 1, 1900?

10. January 1, 1900, a man gave his note for $200 payable July 1, 1902, with interest at 6% from date. How much money will it take to redeem his note when due?

Find the interest on the following notes.

11.

> CHICAGO, IL, March 1, 1900
>
> Six months after date, I promise to pay John Jones, or order, three hundred dollars, with interest at 6% from date.
>
> $300   R. G. SMITH

12.

CLEVELAND, OHIO, May 1, 1900

November 1, 1901, I promise to pay the First National Bank of Cleveland, Ohio, four hundred dollars, with interest at 6% per annum from date.

$400   LEROY HALL

The following is a copy of a bill of goods.

CHICAGO, IL, Oct. 1, 1900

James Ellsworth,

Bought of Werner Pub, Co.

4	Books	@ .25	$1	
3	Tablets	@ .10		30
4	Pencils	@ .05		20
		Total	1	50

> **NOTE**
> Teacher, explain the use of bills.

Make bills for the following:

13. Jan. 1, 1901. B.H. McGrew bought of M.A. Manning:

20 lbs. of sugar @5¢.

20 lbs. of flour @2½¢.

30 lbs. of crackers @8¢.

14. Feb. 15, 19___. A. Baldwin bought of S. Backey & Co.:

100 lbs. of flour @ 2½¢.

15 lbs. of sugar @6¢.

12½ lbs. of prunes @8¢.

10 bu. of potatoes @40¢.

DRILL

Find the interest on:

1. $200 at 6% for 2½ years.
2. $400 at 10% for 6 months.
3. $100 at 6% for 6 months.
4. $100 at 6% for 2 months.
5. $100 at 6% for 2 years and 2 months.
6. $500 at 6% for 2 years and 4 months.

How many board feet in:

7. 5 boards 12 inches wide and 12 feet long?
8. 5 boards 12 inches wide and 16 feet long?

> **NOTE**
> A 16-foot board has 1/3 more board feet in it than a 12-foot board of the same width.

9. How many more board feet would an 18-foot board have than a 12-foot board of the same width?

> **NOTE**
> Another way of estimating board measure is to find out what part of 12 inches the width of the board is, and multiply this by length of the board in feet. For example, how many board feet in a board 9 inches wide and 16 feet long?
>
> $$9 \text{ is } 3/4 \text{ of } 12$$
> $$16 \times 3/4 = 12$$

10. How many board feet in:

    3 boards 6 inches wide and 14 feet long?
    5 boards 10 inches wide and 18 feet long?
    10 boards 3 inches wide and 8 feet long?

✓ Add the following *at sight*.

10	11	11	12	12	13	13	14	14 = ___
11	11	12	12	12	13	14	14	15 = ___
11	12	12	13	13	14	14	15	15 = ___
12	12	13	13	14	15	15	15	16 = ___
12	13	13	14	14	15	15	16	16 = ___
13	13	14	14	15	15	16	16	17 = ___
13	14	14	15	15	16	16	17	17 = ___
14	14	15	15	16	16	17	17	18 = ___
+14	+15	+15	+16	+16	+17	+17	+18	+18 = ___

✓ 1. $35 - 15 \times 3 : 12 =$ ___

2. $40 : 8 \times 9 : 15 =$ ___

3. $32 - 17 \times 3 : 9 =$ ___

4. $21 + 14 : 7 \times 11 =$ ___

5. $35 - 8 : 9 \times 12 =$ ___

6. $42 + 7 : 7 \times 9 =$ ___

7. $12\frac{1}{2}\% \text{ of } 80 \times 9 + 10 =$ ___

8. $33\frac{1}{3}\% \text{ of } 60 - 5 + 16 =$ ___

9. $5\% \text{ of } 100 \times 8 : 20 =$ ___

10. $16\frac{2}{3}\% \text{ of } 36 \times 9 - 4 =$ ___

11. $20\% \text{ of } 45 \times 8 - 12 =$ ___

12. $25\% \text{ of } 40 \times 12 - 60 =$ ___

✓ 13. $2 - 1\frac{1}{2}$ = ____

14. $3 - 1\frac{1}{4}$ = ____

15. $5 - 2\frac{1}{2}$ = ____

16. $6 - 3\frac{1}{3}$ = ____

17. $2\frac{1}{2} - 1\frac{1}{4}$ = ____

18. $4\frac{1}{3} - \frac{2}{3}$ = ____

19. $5\frac{1}{2} - 3\frac{1}{2}$ = ____

20. $20\frac{1}{2} - 5\frac{1}{4}$ = ____

21. $16\frac{2}{3} - 8\frac{1}{3}$ = ____

22. $25 - 12\frac{1}{2}$ = ____

23. $3 + 4\frac{1}{4} - 1\frac{1}{2}$ = ____

24. $5 + 4\frac{1}{2} - 8\frac{3}{4}$ = ____

25. $6 + 8\frac{3}{4} - 9\frac{1}{2}$ = ____

26. $7 + 5\frac{1}{2} - 1\frac{1}{4}$ = ____

27. $8 + 3\frac{2}{3} - 4\frac{1}{3}$ = ____

REVIEW 8

1. How many board feet in a load of sheeting 16 feet long, the total width of the boards being 720 inches?

2. How many feet in a load of lumber containing 100 scantlings, 2 inches by 4 inches, and 12 feet long?

3. What is the weight of such a load, estimating 2½ pounds to the foot?

4. A man wishes to build a fence around a rectangular field 480 feet long and 320 feet wide. How many posts will it take if they are set 8 feet apart?

5. What will it cost to have the post holes dug if he hires a man at 10 cents an hour, and the man digs 10 holes an hour?

6. How many board feet are there in the posts if each post is 4 inches by 4 inches and 6 feet long?

7. What will the posts cost at $20 per thousand?

8. What length boards would you buy for this fence? It is an ordinary fence 3 boards high with the boards running horizontally.

9. How many such boards, 6 inches wide, would be necessary to make the fence?

10. What will the board cost at $20 per thousand?

11. How many pounds of ten-penny wire nails will it take, allowing 7 nails to the board and 70 nails to the pound?

12. What will the nails cost at 3 cents per pound?

13. How many loads of lumber will it take for this fence if 800 feet are hauled at a load?

14. What will be the cost of hauling at sixty cents a load?

15. If it takes 2 men 2 days to build this fence, working 10 hours a day, what will their labor be worth at 10 cents an hour?

16. What is the total cost of fencing this field?

17. A party of missionaries, bound for Cape Town, South Africa, by way of Liverpool, England, left New York at 9 am Wednesday, August 1, 1900. They made the trip to Liverpool in 10 days and 3 hours. After a delay of 2 days in Liverpool, they took ship for Cape Town, arriving there in 26 days and 6 hours from Liverpool. Upon what day of the week did they arrive at Liverpool? How many days and hours did it take to make the trip from New York to Cape Town?

18. Upon what day of the month did they arrive in Cape Town?

19. Upon what day of the week did they arrive there?

20. What was the time of the day when they reached Liverpool? Cape Town?

The following is a typical canvasser's report for 1 month.

	No. of Exhibitors	Orders Taken	Books Delivered & Cash Sales	No. of hours worked
Monday	20	16	2	10
Tuesday	24	18	3	10½
Wednesday	16	12		7
Thursday	19	14	4	10
Friday	15	12		8
Monday	24	20	6	11
Tuesday	17	12	11	10
Wednesday	21	15	8	11
Thursday	24	20	3	11½
Friday	18	12	5	8
Monday	10	6	28	11
Tuesday	12	9	21	10
Wednesday	25	20	4	10
Thursday	10	7	1	5
Friday	20	15		8
Monday	4	2	82	11½
Tuesday	6	4	41	10
Wednesday	4	3	32	11½
Thursday	3	2	16	10
Friday	2	2	9	10
Total				

21. Find the total of each column.

22. What percent of orders were taken from the total number of exhibitions?

23. If all the orders taken were delivered, how many cash sales were made?

24. If the book canvassed for sold for $1, how much money did the agent receive from his subscribers?

25. If his commission was 25%, how much money did he make?

26. How much should he send to the publishing house he was working for?

27. How much tithe should he pay?

28. How much did he make an hour?

29. If he paid $14 for board and room, $9 for other expenses, and then gave 50% of the remainder for missionary work and other good works, how much did he still have?

30. Find the interest on the following note.

> LINCOLN, NEB., May 1, 1900
>
> September 1, 1901, I promise to pay the First National Bank of Lincoln, Neb., two hundred dollars, with interest at 6% per annum from date.
>
> $200   T. J. Merryman

31. Copy and find the amounts due on the following account.

H. A. Jones,
          In account with R. W. Brown.

1900

Jan	2	To 100 lbs. flour	@ $.02½		
	2	To 20 lbs. rice	@ $.08		
	15	To 16 cans of peaches	@ $.12½		
	15	To 12 lbs. prunes	@ $.08⅓		
Feb	5	To 5 gal. gasoline	@ $.12		
	12	To 16 lbs. sugar	@ $.06¼		
	13	To 8 cans of tomato	@ $.09		
			Amount Due		

DRILL

COMMON FRACTIONS TO DECIMAL

1. Reduce ³⁄₅ to a decimal.

> **NOTE**
> ³⁄₅ = ⁶⁄₁₀ = .6

Reduce to decimals:

2. ⁴⁄₅ = _____   8. ³⁄₅₀ = _____

3. ¹⁄₂₅ = _____   9. ¹²⁄₂₆ = _____

4. ⁶⁄₂₀ = _____   10. ⁶⁄₈ = _____

5. ¾ = _____   11. ⁹⁄₂₀ = _____

6. ⁴⁄₁₀ = _____   12. ¹⁵⁄₂₀ = _____

7. ⁴⁄₄₀ = _____   13. ⁷⁄₂₅ = _____

Express decimally:

14. $7³⁄₅ = _____   18. $42⁴⁄₂₀ = _____

15. $12²⁄₁₀ = _____   19. $31⁶⁄₂₅ = _____

16. $12¾ = _____   20. $75½ = _____

17. $66²⁄₄ = _____   21. $21⁴⁄₄₀ = _____

Express as decimals:

22. 8 tenths.

23. 36 and 42 thousandths.

24. 125 thousandths.

25. 7 and 4 thousandths.

26. Four hundred and eight tenths.

27. Four hundred and eight thousandths.

28. Seventy dollars and 1 cent.

29. Fifty four dollars and 3 mills.

30. One hundred dollars and six and one-half cents.

✓ Add the following *at sight.*

6	7	8	9	10	11	12	13	14 = ___
10	11	11	12	12	13	13	14	14 = ___
11	11	12	12	13	13	14	14	15 = ___
11	12	12	13	13	14	14	15	15 = ___
12	12	13	13	14	14	15	15	16 = ___
12	13	13	14	14	15	15	16	16 = ___
13	13	14	14	15	15	16	16	17 = ___
13	14	14	15	15	16	16	17	17 = ___
+14	+14	+15	+15	+16	+16	+17	+17	+18 = ___

🔊 ORAL EXERCISES

✓ 1. 20 – 14 x 9 – 4 : 2 = _____
2. 30 – 18 x 7 + 6 : 3 = _____
3. 40 – 29 x 8 + 2 : 9 = _____
4. 32 – 16 x 2 + 3 : 7 = _____
5. 25 – 14 x 9 + 1 : 10 = _____
6. 73 – 13 : 6 x 9½ = _____
7. 42 – 12 : 15 x 8½ = _____
8. 34 – 14 x 2 + 16 = _____
9. 20 x 3 + 15 : 3 = _____
10. 75 – 19 : 7 x 4 = _____

✓ 11. 1 percent of 100 = _____
12. 1 percent of 200 = _____
13. 2 percent of 200 = _____
14. 3 percent of 200 = _____
15. 4 percent of 100 = _____
16. 5 percent of 200 = _____
17. 3% of 300 = _____
18. 4% of 400 = _____
19. 5% of 100 = _____
20. 5% of 500 = _____
21. 6% of 100 = _____
22. 6% of 600 = _____

✓ Bring in five problems based on physiology.

NUTRITIVE VALUE OF FOODS

Find the total nutritive value of the foods named below. This is done by adding the several elements found in the food. Find the amount of water and waste matter in each food. If the pupils desire to know the exact amount of water or waste, the teacher should give them the information which can be found in any first-class cookbook or textbook on physiology.

Read and add at sight. When thoroughly learned, the pupils may be allowed to write the results in the blank space.

NOTE

Children can be made as familiar with the nutritive value of the common foods as they are with the number of quarters, dimes, and nickels in a dollar; or the number of inches in a foot or rods in one mile. If children could have a working knowledge of the food elements and their proportion in the foods, they could intelligently make proper combinations; for example, they would know how to combine albumin, starch, and fat together in their proper proportions. While at the table, they would know at a glance the food value of an apple, walnut, bread, etc. It is surprising that there is so much ignorance among educated people on this subject. There are learned men in schools and colleges who can discourse upon the heavenly bodies and things upon the earth and in it, who do not know the food elements and their value in a peach, banana, grape, or chestnut. If children should be taught this subject arithmetically, and thoroughly drilled, they would have an intelligent and sure foundation for *true health reform*, and they will greatly enjoy the study and drill.

	Albominous	Carbonaceous	Sugar	Fat	Salt	Pectose	Free Acid	Total Value	Water & Cellulose
Whole wheat bread	8.7	60	4.	6.	3.			81.7	18.3
White bread	5.3	46.	2.3	.8	.5			54.9	45.1
Rye bread	9.	67.5	3.5	1.	2.				
Macaroni	9.	76.5		.3	.8				
Oat meal	15.	67.		7.	2.				
Corn meal	9.7	69.		4.					
Apple	.4		7.2		.5	4.8	.8		
Banana	1.9	32.2			.6	1.			
Grape	.6		14.3		.5	2.	2.2		
Pear	.4		8.2		.3	3.3	.2		
Peach	.7		4.5		.7	7.1	.9		
Strawberry	1.1		6.3	.5	.8	.5	.9		
Dried prune	2.3	13.6	44.4		1.4	4.3	2.7		
Dried apple	1.3	12.1	43.6		1.6	4.8	3.6	67.	33.
Dried raisin	2.4	7.5	54.6		1.2		.6		
Dried fig	4.		49.8		2.9				
Dried date	9.		58.						
Chestnut	14.6	69.			2.4	3.3			
Walnut	15.8	13		57.4	2.				
Hazelnut	17.4	7.2		62.6	3.				
Peanut	28.3	1.8		46.2	3.3				
Carrot	1.2	9.2			.3	1.			
Cabbage	4.	10.4	1.2	.9	1.6				
Celery	1.5	11.	.8	.4	.8				
Turnip	1.5	3.	.2		.7				
Lettuce	1.4	2.2			.3	1.			
Potato	2.2	21.	.2		1.				
Cucumber	1.2	1.4	1.		.4				
Tomato	1.6		2.5	.3	.6		1.8		
Green peas	6.4	12			.5	.8			
Dried peas	25.2	57.2			.2	2.9			
Bean	23.2	59.4			2.1	3.3			
String beans	2.2	5.5	1.2	.1	.6				

	Albominous	Carbonaceous	Sugar	Fat	Salt	Pectose	Free Acid	Total Value	Water & Cellulose
Lentils	25.9	53		1.9	3.				
Milk	4.1		5.2	3.9	.8				
Cream	2.7		2.8	26.7	1.8				
Lean beef	19.3			3.6	5.1				
Fish	18.1			2.9	1.				
Entire egg	14.			10.5	1.5				
Yolk of egg	16			30.7	1.3				
White of egg	20.4					1.6			

The amount of water-free food required by the average man is twenty ounces for one day. The food elements required are albuminouse, 2.8 ounces; fats, 1.2 ounces; carbonaceous, 16 ounces. The proportion is albominous, 7 parts; fats, 3 parts; carbonaceous (starch, sugar, and dextrine), 40 parts. If a person eats two meals a day, this amount can be equally divided for each meal.

1. How many ounces of water-free food would be eaten for dinner? How many ounces of each food element?

2. How many ounces would be eaten by a person in one week? How many ounces of each food element?

3. How many ounces of water-free food would be eaten in 4 weeks? How many pounds would this be?

4. How many ounces of fats would be eaten in 28 days?

5. How many ounces of albuminous substance (such as gluten) would be eaten in 28 days?

6. How many ounces of carbonaceous foods would be eaten in 28 days?

7. What is the difference in food value between potatoes and turnips?

8. What is the difference between lean beef and hazelnuts?

9. What is the difference between lean beef and beans? lentils?

10. Twenty five cents buy 5 pounds of beans; 75 cents pay for how many pounds of beans?

11. If 40 cents pay for 4 pecks of tomatoes, $1.20 pay for _____.

12. If $1 pays for 10 pounds of dried prunes, 30 cents pay for _____.

13. In the year 1900, in 5 American cities, 16000 children became nervous wrecks while attending school. How many did it average to each city?

1. America was discovered in 1492; Martin Luther was born in 1483. How old was Luther when America was discovered?

2. The Great Reformation began in 1521. How old was Luther? How long was it after the discovery of America? How long was it before the settlement of Massachusetts which was begun 1620? How long was it before the settlement of Jamestown in 1607? What was the age of Protestantism in 1607? 1620?

3. First German translation of the Bible in 1521; Wycliffe's English translation was made about 1370. How many years' difference in these translations?

4. 1620 means what? 1492? 1521? 1483?

5. Harvard, the first college in America, was established in 1636. How long was this after the first settlement in New England?

6. The sun was darkened in 1780. How long was that after the Declaraton of Independence?

7. 1798 ended the 1260 years of Revelation 12. How long was it after the darkening of the sun?

8. The falling of the stars was 53 years after the dark day, when was it?

9. Slavery was introduced into America in 1619. It was abolished in 1863. How many years was it allowed?

10. Seven years after the dark day, the Constitution of the United States was adopted. What was the date? How long was this before the close of the 1260 years?

11. The first telegraphic message was sent in 1844. The first volume printed in America was in 1640. What was the difference?

12. How many years since AD 1844? since AD 1833? since AD 1798? AD 1787? AD 1780? AD 1776? AD 1763? AD 538? BC 457?

✓ **EXERCISE IN RAPID THINKING**

State the significance of the following numbers.

12, 128, 32, 3, 144, 160, 9, 320, 60, 24, 16, 4, 8, 10, 25, 2000, 2, 36, 80, etc.

When 12 is written on the board by the teacher, the pupil says, "the number of inches in one foot;" 25, "the number of cents in a quarter of a dollar.

✓ Add the following *at sight*.

11	11	12	12	13	13	14	14	15 = ___
11	12	12	13	13	14	14	15	15 = ___
12	12	13	13	14	14	15	15	16 = ___
12	13	13	14	14	15	15	16	16 = ___
13	13	14	14	15	16	16	16	17 = ___
13	14	14	15	15	16	16	17	17 = ___
14	14	15	15	16	16	17	17	18 = ___
14	15	15	16	16	17	17	18	18 = ___
+15	+15	+16	+16	+17	+17	+18	+18	+19 = ___

✓ Add the following decimals, using a pencil.

1.	3.14	2.	4.07	3.	34.75	4.	45.001
	124.03		3.12		2.09		3.28
	7.4		5.75		1.79		.075
+	5.16	+	8.03	+	3.06	+	1.73

✓ Subtract the following, using a pencil.

5.	348.75	7.	41.42	9.	263.02	11.	35.01
–	116.31	–	31.16	–	131.81	–	19.07

6.	284.65	8.	52.34	10.	42.002	12.	431.16
–	134.31	–	48.17	–	16.001	–	216.08

13. In 5 cities of the United States, there were 16000 little children who became broken down in their nervous system, in one year, how many did it average to each city?

UNITED STATES MONEY

10 mills (m.)	=	1 cent (ct. or ¢)
10 cents	=	1 dime (d.)
10 dimes	=	1 dollar ($)
10 dollars	=	1 eagle (E.)

1. Is there a piece of money called a mill? Name all the pieces of United States money you know of—copper, nickel, silver, gold, and paper.

2. How many cents make 3 dimes? 7 dimes?

3. How many nickels in 3 dimes? 9 dimes?

4. 3 nickels and _____ dimes = 1 quarter

5. 6 nickels and _____ dimes = 1 half dollar

6. How many nickels in 75 cents?

7. A boy bought 8 pounds of rice at 7 cents a pound and gave the merchant a half dollar and a quarter. How much change did he receive?

8. A man gave a merchant a ten-dollar bill in payment for a coat and vest which cost $7.25. What 3 pieces of money did he receive in change?

9. What part of a dollar are:

6¼¢? = _____	33⅓¢? = _____
8⅓¢? = _____	40¢? = _____
10¢? = _____	50¢? = _____
12½¢? = _____	66⅔¢? = _____
16⅔¢? = _____	75¢? = _____
20¢? = _____	80¢? = _____
25¢? = _____	90¢? = _____

LESSON 37

LIQUID MEASURE

> 4 gills (gi.) = 1 pint (pt.)
> 2 pints = 1 quart (qt.)
> 4 quarts = 1 gallon (gal.)

1. How many pints in 1 gallon? in 3 gallons?

2. How many quarts in 4 gallons?

3. If each chamber of the heart holds 1 gill, how many pints of blood does the heart hold?

4. If 6 quarts of milk cost 24¢, what will 1 pint cost?

5. At 5¢ a quart, what will 5 pints of milk cost?

6. If 6 quarts of air are breathed each minute, how many gallons are breathed in 1 hour? If a lamp burns 1 pint of oil each day, what will the oil cost for 24 days at 12¢ a gallon?

7. How many 2-quart jars can be filled from 16 gallons of grape juice?

8. From a barrel containing 30 gallons of rain water, 20 gallons were dipped out. Later, the remainder was measured, and it was found that there were only 34 quarts. How much of it had evaporated?

DRY MEASURE

> 2 pints = 1 quart (qt.)
> 8 quarts = 1 peck (pk.)
> 4 pecks = 1 bushel (bu.)

1. How many quarts of beans can be sold from 3 pecks?
2. At the rate of 5¢ a quart, what are beans worth a bushel?
3. A grocer bought a bushel of beans for $1.25 and sold them for 35¢ a peck. Did he make or lose? and how much?
4. From 32 quarts of strawberries, 21 quarts were canned. How many quarts were lost in shrinkage?
5. A grocer bought 10 bushels of potatoes at 30 cents a bushel and sold them for 10 cents a peck. How much did he make on the 10 bushels?
6. A teamster feeds his horses 3 pecks of oats a day. How long will 15 bushels of oats last him? What does it cost him each day when oats are worth 32 cents a bushel?

LESSON 39
AVOIRDUPOIS WEIGHT

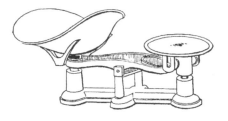

> 16 ounces (oz.) = 1 pound (lb.)
> 2000 pounds = 1 ton (T.)

1. How many ounces are there in 1 pound? in 2 pounds? in 3½ pounds? in 5 pounds?

1 oz.   2 oz.   4 oz.   8 oz.   16 oz.   2 lbs.   4 lbs.

2. Which weight in the above picture equals ¼ of a pound? Which 2 weights equal ¾ of a pound?

3. If a grocer wishes to weigh ⅛ of a pound, which of the above weights would he use?

4. Which ones will he use in weighing ⅜ of a pound? ⁹⁄₁₆? ¾? ⁷⁄₁₆? ⅞? ⅝?

What weights would be used in weighing out:

5. 10 cents' worth of prunes at 8¢ a pound?

6. 15 cents' worth of prunes at 8¢ a pound?

7. 15 cents' worth of sugar at 6¢ a pound?

8. 10 cents' worth of flour at 2½¢ a pound?

9. 10 cents' worth of oatmeal at 4¢ a pound?

10. 25 cents' worth of peas at 5¢ a pound?

11. 5 cents' worth of nails at 4¢ a pound?

12. Could you use the above weights in weighing 1⅕ lbs?

13. How many pounds of rice could you get for 15¢ at 7¢ per pound?

Give the weight of the following articles at the prices given:

14. 25 cents' worth of sugar at 6¼¢.

15. 25 cents' worth of rice at 8⅓¢.

16. 10 cents' worth of flour at 3¢.

17. 25 cents' worth of zwieback at 8¢.

18. 25 cents' worth of raisins at 9¢.

19. 1 dollar's worth of apricots at 12½¢.

20. 1 dollar's worth of sugar at 6¼¢.

21. 1 dollar's worth of fruit crackers at 12¢.

22. 1 dollar's worth of California peaches at 15¢.

23. 1 dollar's worth of nails at 4¢.

Give the cost of:

24. 500 pounds of coal at $8 per ton.

25. 200 pounds of bran at $16 per ton.

26. 400 pounds of hay at $10 per ton.

27. 300 pounds of coal at $8 per ton.

28. 300 pounds of bran at $16 per ton.

29. 300 pounds of hay at $10 per ton.

30. 700 pounds of coal at $7 per ton.

31. 500 pounds of timothy at $12 per ton.

32. 600 pounds of clover at $12 per ton.

LESSON

40 LINEAR MEASURE

12 inches (in.)	=	1 foot (ft.)
3 feet	=	1 yard (yd.)
5½ yards	=	1 rod (rd.)
320 rods	=	1 mile (mi.)
5280 feet	=	1 mile

1. How many inches are there in 2½ feet?

2. If you step 2½ feet at each step, how many yards will you go in taking 30 steps?

3. From a piece of cloth 12 yards long, 4½ feet were cut, how many feet remained?

4. How much will 10 yards of garden hose cost at 8¢ a foot?

5. How much will 60 inches of cloth cost at 30¢ a yard?

6. How many yards in 20 rods?

7. How many feet in 2 rods?

8. How many rods in ¼ of a mile? in ¹⁄₁₆?

9. How many rods of fencing will it take to enclose a field ½ mile square?

10. A man who steps 30 inches at each step, wishes to measure a distance of 10 yards. How many steps must he take?

- Let each student be required to do some measuring by stepping distances. Learn the number of inches taken in each step.
- Require three original problems from each student.

LESSON 41 · SQUARE MEASURE

144 square inches (sq. in.)	=	1 square foot (sq. ft.)
9 square feet	=	1 square yard (sq. yd.)
30¼ square yards	=	1 square rod (sq. rd.)
160 square rods	=	1 acre (A.)

1. How many inches are there in a surface 1 foot long and 18 inches wide?

2. How many square feet are there in a surface 8 inches wide and 3 feet long?

3. How many square yards in a surface 6 feet wide and 9 foot long? in a surface 15 feet long and 12 feet wide?

4. What will it cost to plaster the ceiling of a room 18 feet long and 15 feet wide at 12¢ a square yard?

5. A roll of wallpaper 8 yards long and 18 inches wide cost 24¢. What was the price per square yard?

6. If a piece of land, 4 rods wide and 10 rods long, yields 35 bushels of potatoes, what is the rate of yield per acre?

7. How many acres of land are there in 1 square mile? What part of an acre are 20 sq. rods? 25 sq. rods? 30 sq. rods? 50 sq. rods?

CUBIC MEASURE

> 1728 cubic inches (cu. in.) = 1 cubic foot (cu. ft.)
> 27 cubic feet = 1 cubic yard (cu. yd.)
> 128 cubic feet = 1 cord

1. How many cubic inches in a block 3 inches by 4 inches by 6 inches?

2. How many cu. in. in a brick 8 inches long, 4 inches wide, and 2½ inches thick?

3. What part of a cord would there be in a pile of wood 8 feet long, 4 feet wide, and 2 feet high?

4. A pile of wood is 8 feet long and 4 feet wide. How high must it be in order to contain a cord?

5. How many cu. ft. of air space in a room 30 feet long, 20 feet wide, and 10 feet high?

6. Allowing 300 cu. ft. of air space to each person, how many persons should live in such a room?

7. What will it cost to dig a cellar 18 feet long, 15 feet wide, and 8 feet deep at 25¢ a cu. yd.?

8. How many brick will it take to lay a wall 20 feet long, 10 feet high, and 18 inches thick, allowing 20 brick to the cu. ft.?

9. A pile of wood 8 feet long, 4 feet wide, and 4 feet high is called a cord. How many cubic feet in a cord?

10. What part of a cord is contained in a pile of wood 8 feet long, 4 feet wide, and 2 feet high?

11. If it is worth 50 cents to saw cord-wood into three lengths, what would it cost to have this pile sawed?

12. Make a drawing of a woodshed 16 feet long, 12 feet wide, and 8 feet to the ceiling. Indicate the position in which cord-wood could be piled to the best advantage. How many cords will the shed hold?

13. What does it cost to fill the shed with hard maple at $5 per solid cord?

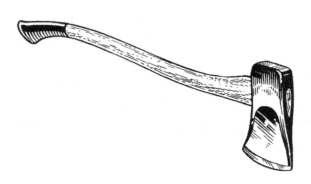

60 seconds (sec.)	=	1 minute (min.)
60 minutes	=	1 hour (hr.)
24 hours	=	1 day (da.)
7 days	=	1 week (wk.)
365 days or 12 calendar months	=	1 common year (yr.)
366 days	=	1 leap year (yr.)
100 years	=	1 century (C.)

1. An aged person's heart beats once every second. How many times will it beat in ¼ of a minute?

2. A middle-aged person's heart beats 5 times every 4 seconds. How many times will it beat in a minute?

3. A 10-year-old child's heart beats 3 times in 4 seconds. How many times will it beat in 1 minute?

4. A passenger train runs 2 miles in 3 minutes. How far does it run in 1 hour?

5. How many minutes does it take the hour hand of a clock to go from IV to XI?

6. A man walked 4 miles in 1 hour and 8 minutes. What was the average time of walking 1 mile?

7. A horse traveled at the rate of 4½ miles an hour, 10 hours every day for 5 days. How far did he travel?

8. How many hours are there in 3/7 of a week?

9. How much does a man make in 1 week who works 10 hours a day at 15¢ an hour?

10. If a man works 10 hours a day on a salary of $12 a week, how much does he get an hour?

11. How much does a man make in 1 year on a salary of $50 per month?
12. How much tithe ought this person to pay every year?
13. How many days from January 1, 1901, to May 15, 1901?
14. How many days are there in each month?
15. How many centuries old is Elijah?
16. How many centuries from the birth of Abraham to the building of Solomon's temple?
17. How many centuries from the flood to the present time?

MULTIPLICATION TABLE

2's

2 x 1 = 2
2 x 2 = 4
2 x 3 = 6
2 x 4 = 8
2 x 5 = 10
2 x 6 = 12
2 x 7 = 14
2 x 8 = 16
2 x 9 = 18
2 x 10 = 20
2 x 11 = 22
2 x 12 = 24

3's

3 x 1 = 3
3 x 2 = 6
3 x 3 = 9
3 x 4 = 12
3 x 5 = 15
3 x 6 = 18
3 x 7 = 21
3 x 8 = 24
3 x 9 = 27
3 x 10 = 30
3 x 11 = 33
3 x 12 = 36

4's

4 x 1 = 4
4 x 2 = 8
4 x 3 = 12
4 x 4 = 16
4 x 5 = 20
4 x 6 = 24
4 x 7 = 28
4 x 8 = 32
4 x 9 = 36
4 x 10 = 40
4 x 11 = 44
4 x 12 = 48

5's

5 x 1 = 5
5 x 2 = 10
5 x 3 = 15
5 x 4 = 20
5 x 5 = 25
5 x 6 = 30
5 x 7 = 35
5 x 8 = 40
5 x 9 = 45
5 x 10 = 50
5 x 11 = 55
5 x 12 = 60

6's

6 x 1 = 6
6 x 2 = 12
6 x 3 = 18
6 x 4 = 24
6 x 5 = 30
6 x 6 = 36
6 x 7 = 42
6 x 8 = 48
6 x 9 = 54
6 x 10 = 60
6 x 11 = 66
6 x 12 = 72

7's

7 x 1 = 7
7 x 2 = 14
7 x 3 = 21
7 x 4 = 28
7 x 5 = 35
7 x 6 = 42
7 x 7 = 49
7 x 8 = 56
7 x 9 = 63
7 x 10 = 70
7 x 11 = 77
7 x 12 = 84

8's

8 x 1 = 8
8 x 2 = 16
8 x 3 = 24
8 x 4 = 32
8 x 5 = 40
8 x 6 = 48
8 x 7 = 56
8 x 8 = 64
8 x 9 = 72
8 x 10 = 80
8 x 11 = 88
8 x 12 = 96

9's

9 x 1 = 9
9 x 2 = 18
9 x 3 = 27
9 x 4 = 36
9 x 5 = 45
9 x 6 = 54
9 x 7 = 63
9 x 8 = 72
9 x 9 = 81
9 x 10 = 90
9 x 11 = 99
9 x 12 = 108

10's

10 x 1 = 10
10 x 2 = 20
10 x 3 = 30
10 x 4 = 40
10 x 5 = 50
10 x 6 = 60
10 x 7 = 70
10 x 8 = 80
10 x 9 = 90
10 x 10 = 100
10 x 11 = 110
10 x 12 = 120

11's

11 x 1 = 11
11 x 2 = 22
11 x 3 = 33
11 x 4 = 44
11 x 5 = 55
11 x 6 = 66
11 x 7 = 77
11 x 8 = 88
11 x 9 = 99
11 x 10 = 110
11 x 11 = 121
11 x 12 = 132

12's

12 x 1 = 12
12 x 2 = 24
12 x 3 = 36
12 x 4 = 48
12 x 5 = 60
12 x 6 = 72
12 x 7 = 84
12 x 8 = 96
12 x 9 = 108
12 x 10 = 120
12 x 11 = 132
12 x 12 = 144

APPENDIX
SUGGESTIONS TO TEACHERS

Teachers should read the preface and study carefully the plan of this book. It is intended that solid work shall be done, and that pupils will have a desire created to see everything from an arithmetical standpoint. It is believed that problems which teach a truth or a useful fact will not detract from arithmetical process. The concrete problems will set forth the principles of arithmetic in a living manner. These problems are followed by drills, which will exercise the pupils upon these numbers until they can use them as automatically as the pianist uses his fingers upon the keyboard.

Teachers should not be satisfied until accuracy is obtained. A failure now and then in giving a result is a weakness that will unfit the pupil for any position where mathematical accuracy is required. A broken link in a chain makes the entire chain useless. Give problems that are not beyond the ability of the students, and hold them upon them until they become proficient.

The teacher should use his own judgment in assigning lessons. Some of the lessons may be too long. Do not rush the pupils over them. Master every lesson before a new lesson is given, unless the class is composed of young pupils who will go through the book more than once. No pupil will be expected to enter the second book until he has met all the requirements of this book. Very young pupils can begin with this book. It can also be used for pupils in advanced grades if they are weak in mental arithmetic.

There is a great advantage in keeping the pupils upon mental arithmetic until they become very proficient in numbers. After they have solved the problems mentally, it will do no harm to allow the problems to be solved with pencil and slate. It is easy to pass from mental to written arithmetic, but not so easy to pass from written to mental arithmetic.

The teacher should always look over the advance lesson with the pupils, to help them to learn how to study. It is the business of the teacher to teach pupils how to study, as well as to hear recitations. Pupils should be urged to do everything that is suggested in the book. For example, when a lesson on measuring is given, the children should use their measuring rules. "Learning by doing," should be the motto. Such work will be of greater value than the catch or puzzling problems so often placed in arithmetics.

Christian education demands that all subject-matter shall be in harmony with God's law. When received into the mind, it should be practiced. It is unchristian to learn truth and not practice it. To teach a pupil to determine the number of cubic feet in a pile of wood by sending him to the pile to measure it, is in accordance with the methods of Christian education, and is making the Scriptures the basis of such work. God's Word requires that we should be intelligent and skillful in all temporal affairs in which we can engage to his glory.

LESSON 1.—These problems are intended to teach truth. Interest the pupils in the subject-matter until they voluntarily bring original problems to the class drawn from objects around them.

DRILL section of Lesson 1.—The numerals used in the Drill section are the same as those employed in Lesson 1. Children delight in adding and subtracting these numbers when they have used them in concrete problems as they have in Lesson 1.

LESSON 2.—Teachers should see that each child is provided with a measuring rule or a tapeline. They will need them from lesson onward.

DRILL section of Lesson 2.—The numerals in this drill are the same as those employed in Lessons 1 and 2. The drills have no new numerals. The children are to be drilled upon the processes until they can rapidly work the problems. The teacher should not be satisfied until the child is able to give the result of 2+2 as readily as to read the figure 4. If they are drilled repeatedly in this way, it will be as easy for a child to determine the result of 4+6 as to read the figure 10.

LESSON 3.—It will be necessary to provide the children with paper and scissors. Each child should have a pair, or access to a pair, as it will be necessary to do considerable paper cutting after this. Every member of the class should do the work required.

DRILL section of Lesson 3.—It would be well to have the pupils bring in a number of original problems from the Bible. They will enjoy reducing the liquid measures used anciently to modern liquid measures.

DRILL section of Lesson 8.—It would be well, when the pupils are first tested on the sight exercises, to record the time required. After considerable drill, notice what advancement has been made in speed.

LESSON 10.—This lesson suggests what may be done in the study of physiology arithmetically. Physiology is a rich field for problems. Physiological charts should be used to illustrate the facts brought out in this lesson.

DRILL section of Lesson 11.—The number exercise can be used many times and will make number work a pleasure.

REVIEW section of Lesson 17.—This exercise should be repeated very often. It is an excellent practice. See that the problems given are not beyond their ability, and then expect every answer to be correct.

LESSON 35.—If the teacher will place the importance of the nutritive value of foods before the children, they will take a keen interest in it and will learn the food elements and their proportions very readily. It would be well to have them go to the board, or to take their slates, and as the different foods are mentioned to them, let them write down the different elements and the amount, until they are very familiar with the more ordinary articles of food. Many problems can be made based upon this information.

DRILL section of Lesson 35.—Exercise in rapid thinking. This is a most excellent plan and should be repeated often. The pupils will enjoy the exercise of running rapidly over all the tables that they have learned until they catch the proper number.

LESSON 36.—The pupils have had many problems embracing the table of United States money and should now be required to learn the table until they can recite it without the slightest hesitation.

LESSONS 37 to 43.—What has been said of Lesson 36 is true of these lessons.

MULTIPLICATION TABLE.—The combination of numbers made in the multiplication tables from the 2's to the 12's has been used throughout the entire book, first in the concrete form, and then in the drill; and the pupil should be so familiar with the combinations as to be able to give them forward or backward, or begin anywhere in the table and go either way.